C0-ALU-548

Zoophysiology Volume 16

Pierre Bouverot

Adaptation to Altitude-Hypoxia in Vertebrates

Chapter 6 Written in Collaboration with C. Leray

With 43 Figures

Springer-Verlag
Berlin Heidelberg New York Tokyo 1985

Prof. Pierre Bouverot
Laboratoire de Physiologie Respiratoire
associé à l'Université Louis Pasteur
Centre National de la Recherche Scientifique
23, rue Becquerel, F-67087 Strasbourg

ISBN 3-540-13602-9 Springer-Verlag Berlin Heidelberg New York Tokyo
ISBN 0-387-13602-9 Springer-Verlag New York Heidelberg Berlin Tokyo

Library of Congress Cataloging in Publication Data. Bouverot, P. (Pierre), 1924– Adaptation to altitude-hypoxia in vertebrates. (Zoophsiology; v. 16) Bibliography: p. Includes index. 1. Altitude, Influence of. 2. Anoxemia. 3. Adaptation (Physiology). 4. Vertebrates – Physiology. I. Title. II. Series. QP82.2.A4B68 1985 596'.0191 84-14131 ISBN 0-387-13602-9 (U.S.)

Printed in Germany

Typesetting, printing and bookbinding: Brühlsche Universitätsdruckerei, Giessen
2131/3130-543210

Preface

Adaptation to altitude hypoxia is characterized by a variety of functional changes which collectively facilitate oxygen transport from the ambient medium to the cells of the body. All of these changes can be seen at one time or another in the course of hypoxic exposure. Yet, as already stressed (Hannon and Vogel, 1977), an examination of the literature gives only a sketchy and often conflicting picture of the exact nature of these changes and how they interact as a function of exposure duration. This is partly because of the limited number of variables explored in a given study, but it is also attributable to differences in experimental design, differences among species in susceptibility to hypoxia, nonstandardized experimental conditions, lack of proper control of physical (e.g., temperature) and physiological variables (e.g., body mass), failure to take measurements at key periods of exposure, and gaps in knowledge about some fundamental mechanisms. Furthermore the available data on animals native to high altitude are meager and/or inconclusive. Extensive further work under well-controlled experimental conditions is required before a detailed picture can be made.

Nevertheless, it has been a guiding principle in the preparation of this monograph rather to summarize the vastly dispersed material that constitutes the comparative physiology of adaptation to high altitude into a coherent picture, than to provide a comprehensive survey of the field. The data from the literature have consequently been interpreted from the point of view of the author, who is a respiratory physiologist.

Accordingly, in the first two chapters I have attempted to introduce, at a general level, the major topics (altitude hypoxia, adaptations, respiratory problems), the diversity among vertebrates and the difficulties in experimental comparative respiratory physiology. I feel that this information is useful for the reader who desires a critical appreciation of many of the experiments that are described. The other chapters, each reasonably complete in itself, focus on particular steps, convective or diffussive, in the O_2 transport into and within the body. In collaboration with C. Leray, I have not hesitated to go to the cellular level, and to look at possible biochemical adaptations that may occur at high altitude. This appears to us an area of investigation that is still in its infancy.

To avoid a cumbersome bibliography (4000 references on the effects of altitude hypoxia were already available 15 years ago; see Wulff et al., 1968), review articles and books have sometimes been cited instead of original papers. We apologize to all contributors not directly quoted.

The book would never have been finished without the help of many people. The editorial help, encouragement and constructive criticisms of Dr. K. Johansen are gratefully acknowledged, as well as the cooperation of the staff of Springer-Verlag in Heidelberg. I should especially like to acknowledge the support of Dr. P. Dejours, director of the Laboratoire de Physiologie Respiratoire, Centre National de la Recherche Scientifique. Throughout the years, the work of this laboratory has centered on some of the subjects covered in the monograph. Special thanks are due to Mrs. M. L. Harmelin (†) for her skilful assistance in the preparation of illustrations, and to Mrs. J. Brencklé and Miss B. Senart for seeking out many papers and contributions.

I also owe tremendous thanks to Odile Marteau, who has patiently helped me through this project; Odile's support, secretarial professionalism, and constant diligence have made my task a genuine pleasure. Lastly, but importantly, I wish to pay special tribute to my wife Danielle, without whose affectionate encouragement none of this could have been possible.

Strasbourg, Autumn 1984 PIERRE BOUVEROT

Contents

List of Symbols and Subscripts

Subscripts

a	arterial	A	alveolar
b	blood	B	barometric
c	capillary	D	dead space
g	gas	E	expired
t	tissular	EP	end-parabronchial
v	venous	ET	end-tidal
$\bar{v}$	mixed venous	I	inspired
w	water	L	lung
x	substance; e.g., O_2, CO_2, or water vapor	T	tidal

Symbols

C	concentration; e.g., $C_{I_{O_2}}$ = oxygen concentration in the inspired medium
D_x	coefficient of diffusion of substance x
F	fraction of a gas in dry gas phase
f_H	heart frequency
f_R	ventilatory frequency
G	conductance; e.g., $G_{a,\bar{v}_{O_2}} = O_2$ conductance between arterial blood and mixed venous blood. $G_{L_{O_2}}$ = lung diffusing capacity
$\dot{M}_x$	quantity of substance x per unit time; e.g. $\dot{M}_{O_2}$ = oxygen uptake
P	gas pressure; may be partial pressure; e.g. P_B = barometric pressure, $P_{\bar{v}_{O_2}} = O_2$ tension in the mixed venous blood
P_{50}	O_2 partial pressure at which 50% of the hemoglobin is oxygenated
R	respiratory exchange ratio
V	volume of medium; e.g., V_T = tidal volume, V_S = stroke volume
$\dot{V}$	volume per unit time; e.g., $\dot{V}w$ = water flow rate, $\dot{V}g$ = gas flow rate, $\dot{V}b$ = blood flow rate

$\dot{V}/\dot{M}_{O_2}$ volume of medium which must be passed over the gas exchange surface to effect delivery of a unit quantity of O_2, also named $\dot{M}_{O_2}$-specific convection requirement; e.g. $\dot{V}b/\dot{M}_{O_2}$ = blood convection requirement or specific cardiac output

β capacitance coefficient; e.g., βb_{O_2} = capacitance coefficient of oxygen in blood

Chapter 1

General Introduction

In much of this book I will consider responses to high altitude as interchangeable with responses to hypoxia, and explore the mechanisms whereby vertebrate animals, including man, adapt to oxygen-impoverished environments, such as in highland areas or in hypoxic waters. The physiological effects of high altitude, however, cannot be simply attributed to the reduction of partial pressure of oxygen in the ambient medium, air or water, and the consequently reduced oxygenation of the body. In addition to hypoxia, a number of physical, climatic, and nutritional factors may influence the degree and kind of adaptation. These must be taken into account when comparing high- and low-altitude individuals. On the other hand, the concept of adaptation may lead to a conflict of opinion if not properly defined. For these reasons, Sections 1.1 and 1.2 attempt to define "high altitude" and "adaptation" as used in this book, and to provide an overview which may help toward understanding some conflicting data reported in subsequent chapters. Section 1.3 explains how the topic of adaptation to altitude hypoxia will be considered.

1.1 High Altitude

There is no clear distinction between "low" and "high" altitudes. It is commonly accepted with reference to humans living at low altitude that symptoms such as shortness of breath, rapid pulse rate, interrupted sleep, and headaches occur at about 3 km elevation in the majority of newcomers to highland locations (Heath and Williams 1981). Yet these symptoms may start to occur at lower elevations, while changes in some physiological functions such as aerobic capacity (Squires and Buskirk 1982) or night version (McFarland and Evans 1939) become apparent at about 1.2–1.5 km elevation. On the other hand, high-altitude research stations throughout the world are found at elevations ranging approximately from 1.3 to 4.9 km (Table 1.1), illustrating that the term "high altitude" has no precise definition.

Decreases in barometric pressure and in partial pressure of oxygen are the only stressors[1] constantly encountered in mountains. All other stressors examined here vary in intensity and relative importance irregularly at a given altitude

1 According to Adolph (1956, p 18) a stressor is "a specified treatment or circumstance to which animals are exposed ... It comprises all the differences between the situation that may induce modifications and the control situation with which it contrasts"

Table 1.1. The elevation of "high-altitude" research stations (Weihe 1964). It is clear that much depends on the definition of high altitude

Altitude m	Station	Country
4880	Ticlio	Peru
4540	Morococha	Peru
4340	White Mountain (Summit Laboratory)	California
4300	Mt. Evans	Colorado
4200	Cerro de Pasco	Peru
3990	Mina Aguilar	Argentine
3800	White Mountain (Barcroft Laboratory)	California
3454	Jungfraujoch	Switzerland
3090	White Mountain (Crooked Creek Laboratory)	California
3000	Grossglockner	Austria
2900	Cal d'Olen (Instituto Angelo Mosso)	Italy
2220	Westgard Pass	California
2000	Obergurgl	Austria
1600	Denver	Colorado
1500	Salt Lake City	Utah
1260	Jujuy	Argentine

and from one area to another, depending on the latitude, season, diurnal rhythms, and meteorological conditions. Therefore, these characteristics do not help in deciding the physical limits of "high" and "low" altitudes.

1.1.1 Environmental Factors

Low pressure of oxygen, cold and significant diurnal variation in ambient temperature, low humidity, reduced atmospheric absorption of radiant energy, rugged topography and poorly developed soils with marginal availability of certain nutrients are some of the factors inherent to life at high altitude. These factors can be grouped into four categories of stressors: hypoxia, cold, dehydration, malnutrition.

1.1.1.1 Hypoxia

Reduced availability of O_2 due to lowered barometric pressure is the fundamental problem associated with life at high altitude, as Paul Bert's experiments, made partly on himself, and partly on animals, showed conclusively.

A major reason why Paul Bert was able to make this fundamental demonstration was that he equipped his laboratory with decompression chambers. The following quotation taken from his book La Pression Barométrique (1878) describes the favorable action of oxygen in averting the symptoms of decompression:

"I resolved to begin by experimenting on myself. I had already undergone, in my large sheet-iron cylinders, rather considerable decompressions, to the point of experiencing certain discomforts. I then thought of trying the test again, so as to remove the symptoms by breathing a superoxygenated air.

I placed beside me in the apparatus a large rubber bag, containing air whose oxygen content was in proportion to the degree of decompression. ...

Here is an experiment in which in an hour and a quarter I reached a minimum pressure of 248 mm, that is, less than a third of normal pressure, during which experiment I remained 45 min below 400 mm, without having experienced discomfort from the moment when I began to breathe the superoxygenated air regularly. My pulse ... remained from then on at its normal figure; it even dropped towards the end, either because of the long rest in a seated posture, or under the influence of breathing superoxygenated air. Beside me, a sparrow and a rat were very sick, and their temperature dropped several degrees. As for me, far from running any risk, I felt none of the slight discomforts of decompression, nausea, headache, or congestion of the head, nor did I feel any after leaving the apparatus." (Bert, English translation, 1978).

Decompression chambers, also called low-pressure chambers, or hypobaric chambers, in which hypoxia is the environmental variable allow reproduction of the majority of the physiological changes caused by high altitude. They are now widely utilized in many laboratories throughout the world. Numerous hypoxic studies, however, were and still are conducted near sea level without the help of decompression chambers. The rationale for these different approaches is presented in the four subsections that follow.

a) Barometric pressue (P_B) results from the earth's gravitation and attraction on the atmosphere. Theoretically P_B decreases exponentially with the distance from the earth's center. At a given altitude, however, P_B varies with latitude, season, and weather pattern. Hence, some degree of uncertainty is unavoidable in any theoretical evaluation of P_B in relation to altitude.

The dotted line in Fig. 1.1 expresses the P_B vs. altitude relationship based on the early formula proposed by Zuntz et al. (1906) and discussed by Pugh (1957), while the solid line shows the internationally adopted values for average atmospheric conditions in temperate latitudes in accordance with the International Civil Aviation Organization (ICAO 1964; see also Table A.1 in the Appendix). It is clear that the two curves diverge at altitudes above 2 km. This deviation may be serious in physiological evaluations, and stresses the need for direct P_B measurements at the locations under consideration.

From direct measurements in the course of summer mountaineering expeditions and from radiosonde meteorological data there is evidence that actual values fall somewhere between the two curves. On the summit of Mt. Everest, the highest elevation on earth (estimated to be 8848 m), the first direct measurements of P_B were made in the fall of 1981 by the American Medical Research Expedition (West 1982). These measurements agreed well with predictions based on the interpolation of radiosonde balloons data in New Delhi (West and Wagner 1980). They gave a P_B value between 250 and 253 Torr (asterisk in Fig. 1.1), whereas the value predicted from the ICAO curve at the same altitude is 235 Torr, and that from the Zuntz's curve 268 Torr. Hence, the pressure measured at the summit of Mt. Everest turns to fall midway between the two theoretical lines shown in Fig. 1.1. As emphasized by West and Wagner (1980), the few Torr difference between actual and predicted values, as well as the daily variation in P_B due to variable weather conditions, are of physiological importance. The value of P_B, not that of elevation, has primary biological significance. Therefore, the recommendation made by Dill and Evans (1970) was "report altitude if you wish, but always report barometric pressure."

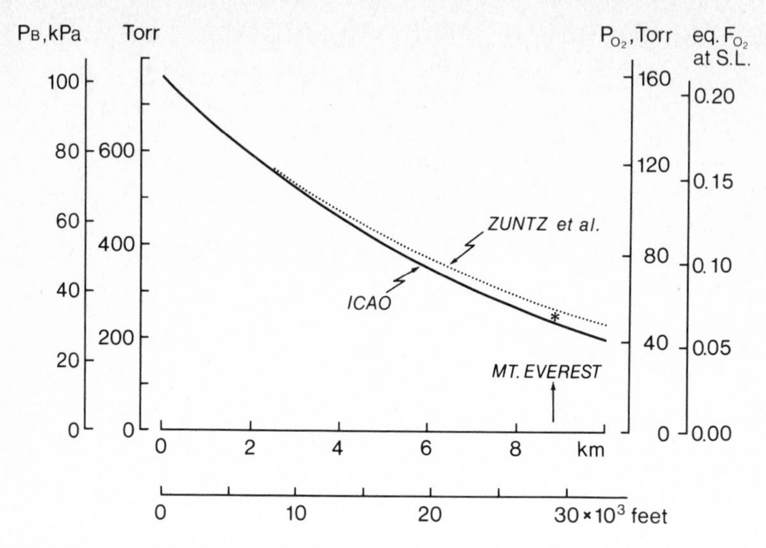

Fig. 1.1. Barometric pressure (PB, *ordinates, left*) as a function of altitude (*abscissa*). *Solid line* derived from the Manual of the ICAO Standard atmosphere (International Civil Aviation Organization 1964); *dotted line* calculated from the equation of Zuntz et al. (1906) as quoted by Pugh (1957; p 592, using a mean temperature of 15 °C). *Asterisk* designates the first direct record obtained at the summit of Mt. Everest (West 1982). *Ordinates on right* indicate the corresponding partial pressures of oxygen in dry air (P_{O_2}) and the fractional concentration of O_2 in normoxic or hypoxic gas mixtures which, *at sea level*, leads to equivalents P_{O_2} values (eq. F_{O_2} at sea level)

b) Partial pressure of oxygen (P_{O_2}) in the dehydrated ambient air relates directly to barometric pressure according to the following equation

$$P_{O_2} = P_B \cdot F_{O_2}, \quad (1.1)$$

where F_{O_2}, the fractional (dimensionless) concentration of oxygen in dry (or dehydrated) gas, remains constant up to an altitude of 110 km (Frisancho 1975). Since dry air contains 20.95% O_2 molecules, $F_{O_2} = 0.2095$ whatever the elevation is. At sea level, therefore, P_{O_2} in the dry air is 159 Torr, assuming that PB is 760 Torr. Both values are halved at about 5.5 km, and reduced by approximately 67% at the summit of Mt. Everest, where P_{O_2} in dry air is 52 Torr, assuming that PB is 250 Torr (Fig. 1.1).

From a practical point of view, it is important to realize that the P_{O_2} value read in Fig. 1.1 (ordinate on right) at given altitude (abscissa) is the maximal possible value as calculated for dry air according to Eq. (1.1). In practice, the ambient air is not dry. Instead it contains more or less water vapor, depending on the relative humidity and temperature. Then the following equation applies:

$$P_{O_2} = (P_B - P_{H_2O})F_{O_2}, \quad (1.2)$$

where P_{H_2O}, the water vapor partial pressure, is independent of the barometric pressure but increases with temperature (Table A.2). In physiological studies,

P_{H_2O} is determined for body temperature (Tb) because the air in contact with the respiratory exchange surfaces is thought to be saturated with water vapor at that temperature. Consequently, the water-saturated inspired air is more hypoxic than the ambient air. For example, since the saturation P_{H_2O} value at 37 °C is 47 Torr (Table A.2), therefore the P_{O_2} value in the air entering the lung of a mountaineer at the summit of Mt. Everest will be (250 - 47) × 0.2095 = 42.5 Torr, a value 20% lower than in the ambient dehydrated air in which P_{O_2} = 52 Torr.

It seems that no consideration has been given yet to the advantage that ectotherm animals may have at high altitude, due to lowered body temperature, which results in lowered vapor pressure of water in the lungs, and therefore higher partial pressure of oxygen.

Equation (1.2) allows theoretical calculation of the value of P_{O_2} in water equilibrated with air at given barometric pressure and temperature. For instance, at 10 °C (P_{H_2O} = 9.2 Torr; Table A.2) and 3 km altitude (PB = 525 Torr), the P_{O_2} value is (525 - 9.2) × 0.2095 = 108 Torr in the air saturated with water vapor at 10 °C, as well as in the aerated water at the same temperature. However, two biological processes, respiration and photosynthesis, act in opposite directions to modify partial pressure of O_2 (and CO_2) in water. Depending on the relative intensity of the photosynthetic O_2 production (which, of course, requires the presence of green plants and light) and the obligatory O_2 consumption by aquatic animals, large spatial and temporal variations in the levels of P_{O_2} (and P_{CO_2}) may occur (Garey and Rahn 1970). Against this background, values of P_{O_2} in aquatic media cannot be predicted, they must be measured, directly.

c) Molar concentrations of oxygen in air (Cg_{O_2}) *and water* (Cw_{O_2}) both decrease in proportion to the decrease of P_{O_2} with ascent to higher elevations, according to the general equation

$$\Delta C_x = \beta_x \cdot \Delta P_x \qquad (1.3)$$

which defines the capacitance coefficient, β_x, of a medium for substance x (O_2 or CO_2) as the change in concentration, ΔC_x, per unit change in partial pressure, ΔP_x. At identical P_{O_2}, however, the O_2 concentration in water is considerably lower than in air (see Fig. 3.1, lower part) owing to the difference between the O_2 capacitance coefficients in the two media (βg_{O_2} and βw_{O_2} in Table A.3).

d) Methods used in hypoxic studies can be distinguished as follows on the basis of Eq. (1.1). First, PB may decrease while F_{O_2} remains constant. This is the hypobaric mode. This situation may be natural, as in field studies in mountains or experiments at high altitude research stations, or artificial when studies are made in low-pressure chambers. These natural and artificial conditions will be referred to as *high-altitude hypoxia* and *hypobaric hypoxia,* respectively. The latter can be continuous or intermittent, depending on the nature of the low-pressure chamber and experimental strategy. In the continuous procedure, the chamber is set to operate continuously at the same level of simulated altitude, automatically adjustable within a few Torr (this implies the existence of a lock and a chamber large enough for allowing the researchers to transit without change in PB). In the intermittent procedure, the depressurized chamber is periodically returned to normal ambient pressure, for instance to clean and feed the animals. Intermittent hypobaric hypoxia repeatedly generates complex environmental stressors: changes in P_B and P_{O_2}, noise, vibrations, decrease in temperature during depressurization and increase during recompression. Continuous hypobaric hypoxia more closely simulates exposure to high-altitude hypoxia, making it easier to distinguish between acute, prolonged, and chronic hypoxia, depending on the time spent at the simulated elevation. Yet continuous hypoxia decreases appetite, particularly in the early phase of hypoxic exposure at (simulated) elevations greater than 3.5 km, a complicating factor which is possibly avoided in intermittent hypoxia.

The second way of decreasing ambient P_{O_2} is to lower F_{O_2} at constant P_B value by premixing the gas to be inspired. This is the normobaric mode, an artificial but easy way of simulating altitude hypoxia. This experimental condition will be referred to as *normobaric hypoxia*. The far right ordinate in Fig. 1.1 shows the values of F_{O_2} which at sea level ($P_B = 760$ Torr) and according to Eq. (1.1) yield the P_{O_2} values encountered in dry air at various elevations.

Some metabolic responses were found to be identical when rats were exposed to either one of these two methods (Russell and Crook 1968).

1.1.1.2 Cold

Much information related to climatic conditions in mountainous regions can be gained from Baker and Little (1976), Baker (1978), Webber (1979), Heath and Williams (1981). For present purposes it suffices to stress that, in general, temperature drops by about 0.6 °C for every 100 m increase in elevation, whatever the latitude is. Thus, at an elevation of 4 km it should theoretically be 12 °C cooler than at 2 km, and 24 °C cooler than at sea level. The actual value, however, depends on latitude. This explains why, at identical elevations, temperature is cooler in the Alps or Himalayas than in equatorial mountains. On the other hand, how temperature influences physiological responses depends on many factors such as the wind velocity, cloudiness, solar radiation, and re-radiation of heat to sky. The two latter factors increase in importance with higher elevation because there is less atmospheric absorption of transmitted radiant energy. As a result, sudden and extreme temperature changes may occur from hour to hour in the course of a day. For that reason the days may be very hot in the tropical Andes, whereas the nights are very cold. These mountains in low latitudes, therefore, experience a well-marked diurnal regime of temperature; in contrast, seasonal changes are small. The converse holds for mountains in higher latitudes.

In subsequent chapters, it should be kept in mind that hypoxia and cold have possibly been interacting factors in many studies made in natural environments at high altitude.

1.1.1.3 Dehydration

Hypohydration, if not dehydration, is an environmental hazard that faces living organisms at high altitudes. Two mechanisms are responsible for increased water loss from the body.

First, the partial pressure of water vapor (P_{H_2O}; Table A.2) and water concentration (C_{H_2O} in Fig. 1.2, left part) decrease in ambient air as a result of decreased

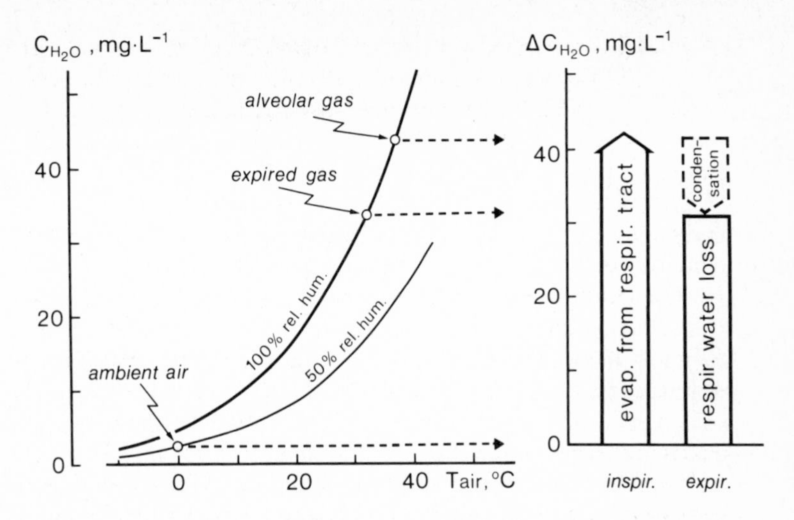

Fig. 1.2. Concentration of water (as vapor; C_{H_2O}) as a function of temperature and relative humidity, and respiratory water exchange (ΔC_{H_2O}) in a cold environment (derived from Walker et al. 1961). As shown in *left* part, C_{H_2O} is very low in ambient air at 0 °C and 50% relative humidity, but increases rapidly with temperature and relative humidity. *During inspiration,* air is progressively warmed and humidified along the respiratory tract; when air enters the lung alveoli at 37 °C and 100% relative humidity, 41 mg · L^{-1} of water have been transferred to it by evaporation from the respiratory mucosa; during that process upper airways have been cooled. *During expiration,* the lowered temperature of upper airways accounts for the cooling of exhaled gas, to about 32 °C in this example, and for the condensation of a certain amount of water (10 mg · L^{-1}, i.e., 25% of the water transferred to inspired air); the expired gas is still saturated at this lower temperature, and the remaining water is lost. This phenomenon, coupled with hypoxic hyperventilation, accounts in part of dehydration at high altitude

temperature at increasing elevations. This phenomenon is independent of barometric pressure, and relates solely to cold and relative humidity. In air-breathers, the passage of air with low water content over the moist surfaces of the airways, coupled with hypoxic hyperventilation (Chap. 3), considerably increase evaporation from the respiratory tract, particularly during muscular activity. From the respiratory water exchange shown in the right part of Fig. 1.2 for a human climber, it may be calculated that about 1.0L of water would be lost through respiration in the course of 10–11 h of climbing with an average pulmonary ventilation of 50 L · min^{-1}. This, in association with the increased evaporation of sweat, accounts for the rapid body dehydration observed during high altitude climbing.

The second mechanism responsible for increased water loss at high elevations relates to the increase in the diffusion coefficient of water vapor, which, as for any gas, is inversely proportional to the barometric pressure (Chap. 2.1.3). Thus, at 0.5 atmosphere when the elevation is about 5.5 km, one would, theoretically, expect the water loss through permeable surfaces to be exactly twice as great as at sea level. Hence the rarefied and arid condition of the atmosphere would favor dehydration of organisms in the absence of special adaptation, either structural or behavioral.

Finally, hypoxia may act independently in reducing thirst, hence water intake, at least in rats (Koob and Annau 1973). The mechanisms of this hypodipsic effects of hypoxia remain to be clarified (Hannon 1981).

1.1.1.4 Malnutrition

Inadequate nutrition is a stressor tied to high altitude. It may affect physiological functions in the organisms making up mountane biological communities, as well as in those exposed to hypobaric hypoxia in a low-pressure chamber.

In highland areas, low soil and air temperatures, together with diurnal and seasonal changes in solar ultraviolet and visible radiation, snow distribution, wind speeds, steepness of slope and soil characteristics make productivity in alpine vegetation low compared to that of the rest of the biosphere. Since green plants are the nutritional base for all members of any biotic community, all animals and humans growing and living at high altitudes may be subject to at least seasonally deficient caloric and nutrient intakes. Yet interaction between nutrition and high-altitude stress are not well known (see Mazess 1975; Baker and Little 1976; Baker 1978; Webber 1979; Hass 1981).

In newcomers to highland areas, particularly at elevations greater than 3.5 km, as well as in subjects or animals exposed to hypobaric hypoxia in a low-pressure chamber, anorexia and hypophagia are known to be most pronounced in the early phase of hypoxic exposure. Underlying mechanisms are still poorly understood (Hannon 1981). The reduced food intake may be responsible for loss of body weight (Fig. 1.3), negative water and nitrogen balances in adults, and may be associated with depressed growth in immature animals. Unfortunately, very

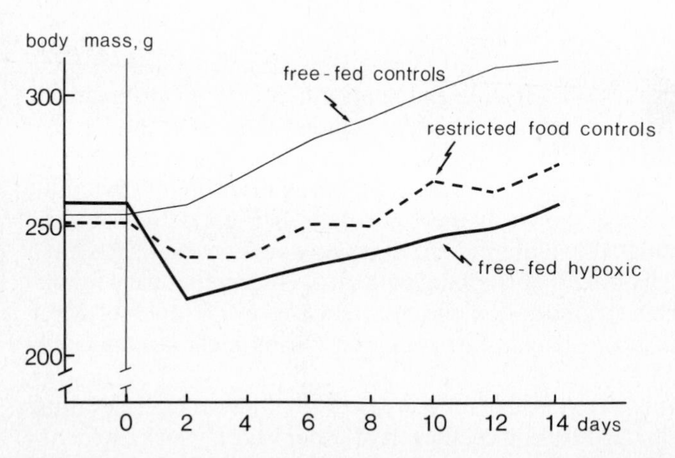

Fig. 1.3. The lower body mass of animals exposed to hypobaric hypoxia appears to be explicable by their loss of appetite. (After Gloster et al. 1972). Three groups of adult male Sprague Dawley rats were studied at 20°–22 °C. All animals were given identical standard pellets and allowed unlimited access to water. Six rats were exposed to hypobaric hypoxia for 2 weeks (P_B = 400 Torr; free-fed hypoxic) with food ad libitum, six acted as a free-fed control group, and the remaining six acted as food-restricted normoxic animals. In these latter, the amount of food given each day corresponded to the weight eaten by the hypoxic animals in the previous 24 h

few studies allow these effects to be ascertained and to be distinguished from those of hypoxia (Gloster et al. 1972; Ettinger and Staddon 1982).

1.1.2 Major Mountain Systems

Figure 1.4 shows schematically that mountains exist on all continents, on many islands, in all latitudes. It has been estimated that about 30% of land surface is above 1 km altitude (Webber 1979). As examined by Pawson and Jest (1978) there are three major high-altitude areas that support sizeable and permanent human and animal populations. These are the Tibetan plateau and Himalayan region in Asia (average altitude 4.5 km, with 15 peaks extending above 7.9 km), the Andes of South America with the cordillera and the high plateaus called altiplano (average altitude around 4 km) and the Ethiopian region in Africa (ranges from 2.4 to 3.7 km with the highest peak at 4.62 km). Among other high-altitude regions are the Alps, Pyrenees, Ural, and Caucasus Mountains in Europe, Atlas Mountains and Kilimanjaro in Africa, Rocky Mountains and Sierra Madre in North America; they do not compare in elevation with the Andes or Himalayas, and the total population from these regions is small.

There are important differences in surface features, latitudinal location, and climatic variables among mountain systems. In Eurasia, the Pyrenees, Alps, Carpathians, Caucasus, Himalayas extend from West to East, paralleling the lines of

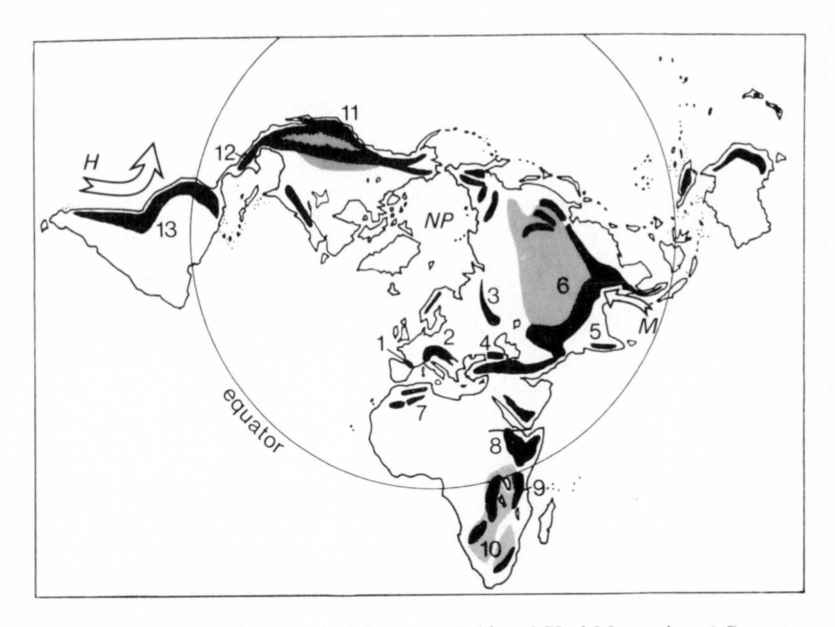

Fig. 1.4. Major mountain systems of the earth. *1* Pyrenees; *2* Alps; *3* Ural Mountains; *4* Caucasus Mountains; *5* Himalayas; *6* Tibetan Plateau and Southern China; *7* Atlas Mountains; *8* Highland Plateau of Ethiopia; *9* Highlands of Kenya; *10* Basutoland; *11* Rocky Mountains; *12* Sierra Madre; *13* Andes. *H* cold Humboldt current, which flows along the Peruvian Pacific coast; *M* wet summer monsoon which originates from the Indian Ocean

latitude. They coincide in general with climatic limits, and intensify them. In contrast, the Rocky Mountains, Sierra Madre, and Andes extend from North to South, offering a great variety of climatic conditions. The direction of mountain ranges has an important influence on their attendant climatic effects. In the tropical Andes, for example, only a few peaks are snow-covered, and dryness is a dominant feature. This relates to the climatic influence of the cold Humboldt current which flows along the Peruvian Pacific coast (H in Fig. 1.4). In contrast, the eastern Himalayas, exposed to the summer monsoon originating from the Indian Ocean (M in Fig. 1.4), are covered with snow and ice; the western extremity of the Himalayas, however, and the Tibetan plateau have very low rainfall (Pawson and Jest 1978). Apart from the latitudinal changes in temperature, there are many other variables such as slope orientation, solar radiation, and wind, which are responsible for microclimatic aspects in mountain regions. Clearly, reduced barometric pressure, with the correlated decrease in the partial pressure of oxygen, is the single denominator common to high-altitude conditions.

1.1.3 Animal Life at High Altitudes

Human populations and a great variety of animals belonging to many different phyla have become established in almost all of the mountain ranges of every continent. Detailed accounts of these can be found elsewhere (Baker 1978; Folk 1966; Hesse 1947; Mani 1962, 1974; Monge and Monge 1968; Napier 1972; Pearson 1951; Swan 1961; Walker et al. 1975). Invertebrates will not be considered in this book, although the insect communities above the timberline in the Himalayas are remarkable for the great abundance of individuals (Mani 1962), and salticid jumping spiders live up to 6 km (Swan 1961). A broad categorization of high-altitude vertebrates is offered below, but unfortunately relatively little is known about respiratory processes in animals that have evolved at high altitude. Much of our present knowledge has resulted from studies made on species, mainly man and a few other mammals, translocated from sea level.

1.1.3.1 Lower Vertebrates

a) Amphibians at high altitudes depend on the presence of sufficient moisture. In the humid zones of the Andes, frogs of the genus *Elentherodactylus* are abundant in the snow line at 4.5 km (Hesse 1947); those of the genus *Telmatobius* have developed a totally aquatic life in the high Andean lakes (Monge and Monge 1968; Hutchison et al. 1976). The viviparous *Salamandra atra* ranges up to 3 km in the Alps and Rocky Mountains, while in the Himalayas the green toad, *Bufo viridis*, can be found at 5 km (Hesse 1947).

b) Reptiles: Lizards are more numerous than snakes, and all of the species found above 3 km in the mountains are ovoviviparous, "... a characteristic that makes possible their entry into the cold alpine zone: there would not be enough heat to bring reptilian eggs to complete development, while the viviparous female lizard or snake is able to follow and keep the sun" (quoted from Hesse 1947). Some species are found in the Himalayas at 5.5 km, while some Iguanidae live at about 4.9 km in the Peruvian Andes (Mani 1974). Sceloporine lizards may be found from sea level to 3.35 km in the south-western USA (Snyder and Weathers 1977). The common African skink, *Mabuya varia,* also viviparous, ranges up to 4 km on Kilimanjaro.

c) Fish (e.g. trout) have been raised in lakes and rivers above the timber line; they have reproduced abundantly in lake Titicaca, and in other lakes and rivers up to 4 km in the Andes (Monge and Monge 1968), and up to 3 km in the Western United States (Rahn et al. 1973). In the Alps, fish ascend to about 2.8 km but, in the Asiatic highlands, several species of the genus *Nemachilus* have been reported in springs at 4.7 km (Hesse 1947), while trout are known to occur in lakes at 3–4 km elevation (Zutshi et al. 1972).

1.1.3.2 Birds

It has long been recognized that a few species of birds are able to fly at extreme altitudes. One of the most remarkable is the bar headed goose, *Anser indicus,* which winters at low elevations throughout the Indian continent, and breeds on large lakes in the south-central regions of Asia at elevations as high as 5.5 km. A flock of these birds has been reported as flying above the Himalayan peaks, including Mt. Everest (Swan 1961). In hypobaric chambers of the laboratory, this bird was still able to stand and hold its head erect at simulated altitudes close to 11 km (Black and Tenney 1980). On the other hand, from various observations made in the Rocky Mountain, Andean, and Himalayan regions, Rahn (1977) has listed 21 avian species, representing ten orders, that nest between 4 and 6.5 km. Scavenger birds such as choughs (*Pyrrhocorax graculus*), lammergeiers (*Gyptaetus barbatus*), and vultures (*Gyps himalayensis*) are known to visit the highest camps of Himalayan expeditions (Swan 1961; Napier 1972). Domestic fowl are raised for meat and egg production at an altitude as high as 4 km (Pearson 1951).

1.1.3.3 Mammals

a) Human beings living in mountain systems can be categorized into different populations. At least, lowlanders and highlanders should be distinguished.

The term "lowlanders" refers to subjects who originate from an area of low elevation and take up a more or less prolonged residence at higher elevation for reasons such as tourism, trekking, mountaineering, business, scientific or military purpose, etc. Depending on the duration of their stay at high altitude, a few hours or days, lowlanders endure the effects of either *acute* or *prolonged* hypoxia; after a few weeks or months, they may become acclimatized to *chronic* hypoxia (Sect. 1.2.2.3).

Ascent to high altitude by people of all ages is more commonplace than in the past, a phenomenon which may be of considerable importance (Heath and Williams 1981). For instance, every summer day in the French Alps, some 4000 tourists ascend from 1.0 km (Chamonix) to 3.8 km (Aiguille du Midi) in 20 min by cable car; then they admire the landscape, ski, or climb higher summits, such as Mt. Blanc. The rapidity of such an ascent is comparable to that in a low-pressure chamber, and precludes any chance of adaptation in unhealthy people. The same holds for those trekkers in the Himalayas who fly to the highest possible altitude to start their trek. Due to increased access to high-altitude areas, largely promoted by the development of the travel industry, the number of newcomers at high altitudes is increasing explosively. This could constitute a growing problem of potential maladaptation, acute mountain sickness (headache, nausea, vomiting, insomnia) and pulmonary edema (Chap. 4). The tremendous physical and mental feat of Reinhold Messner and Peter Habeler, who scaled Mt. Everest without the use of supplemental oxygen (Messner 1978), that of Nicolas Jaeger who spent 60 days alone at 6.7 km elevation on Mt. Huascaran (Jaeger 1979) would not have been possible in the absence of exceptional physical capabilities, perseverence in training, and patient acclimatization at much lower altitudes.

The term "highlanders" refers to those who were born and raised at high altitudes. For climatic and historical reasons, there are three major high-altitude areas in the world permanently inhabited by humans (Pawson and Jest 1978): the Himalayas, Ethiopia, and the Andes. According to reliable census information (Baker 1978) the population size above 2.5 km is approximately 4 million in the Himalayan valleys and Tibetan plateaus, 6.5 million in Ethiopia, and 17 million in the Andean region (Bolivia, Columbia, Ecuador, and Peru). The total population from other high-altitude regions (Rocky Mountains, Alps, Atlas, Ural Mountains, Caucasus, etc.) is smaller. This can be attributed to the fact that environmental stresses do not strictly correlate with altitude. For example, the conditions of life at 3 km elevation are far less severe in the Andean region than in the Alps. This explains why, although major studies have been conducted in many mountain areas, the greatest amount of work research has been done in the Andes.

Cities with relatively large populations exist and considerable mining activity goes on in highlands, at elevations as high as 4–4.5 km, in South America. Nearly all highland populations live on an agro-pastoral basis (Baker 1978). Farming and herding imply domestic animals. These are considered next together with wild species.

b) Other mammals: The yak of central Asia (*Poephagus grunniens*) is the only bovid indigenous to high altitudes. Being used up to 5.8 km (Hesse 1947), it is the highest domesticated animal in the world. Its place is filled in the Andean regions by the domestic camelids llama (*Lama glama*) and alpaca (*Lama pacos*). These, in association with two wild species, the guanaco (*Lama guanacoe*) and vicuña (*Vicugna vicugna*) thrived successfully at an altitude as high as 4.8–5.4 km (Pearson 1951; Koford 1956). All of these camelids have been traditionally considered as autochtonous (native) high-altitude animals found in the Andes since prehistoric times. By contrast, Andean sheep and cattle originated from Europe and were introduced during the 16th century (Monge and Monge 1968). A high-altitude deer, the taruca (*Hippocamelus antisensis*) ranges up to 6 km in the Andes (Hochachka et al. 1983). Sheep are raised up to 5 km, but cattle do not thrive above 4 km (Pearson 1951). Wild sheep and goats are also found in other mountain areas. The chamois (*Rupicapra*) is the most familiar European form.

Among the many rodents that thrive at high altitudes, the guinea pig (*Galea musteloide*) is frequently cited as native to the high Andes, where it provides an important source of food for high-altitude populations (Monge and Monge

1968). The chinchilla is also reported to range in the Andes up to 5 km, while the golden-mantled ground squirrel (*Citellus lateralis*) and the subspecies *rufinus* and *sonoriensis* of the deer mouse (*Peromyscus maniculatus*) inhabit wide range of altitudes up to 4 km in the Rocky Mountains (L.R.G. Snyder et al. 1982).

Dogs are common in the Andean region, whereas domestic cats are rarely found above moderate heights (Monge and Monge 1968). A typical alpine mammal, the marmot, ranges to 3 km in the Alps, but 5.5 km in the Himalayas; yet the marmot moves downward to hibernate on the timberline (Hesse 1947). The Himalayas are extremely rich in mammalian fauna. According to Napier (1972), not only the yak (*Bos grunniens*) but also the Tibetan wild sheep (*Ovis ammon hodgsoni*), antelope (*Panthalops hodgsoni*), and gazelle (*Procapara picticaudata*), the fox (*Vulpes vulpes montana*), wolf (*Eanis lupus chanco*), bear (*Ursus arctos isabellinus*), lynx (*Lynx lynx isabellinus*) among other species, are found above 5 km. Virtually nothing is known about the respiratory characteristics of these animals (Bartels et al. 1963). Some few studies have described the effects of high altitude on horses or mules (Riar et al. 1982).

1.2 Adaptations

The term adaptations, as used in this monograph refers to changes in structures and functions of organisms that result from acute, prolonged or chronic changes in environmental factors. Adaptation is a term that cannot be reduced to a single definition without gross oversimplification. The term has numerous meanings in biology, and occasionally leads to semantic problems.

1.2.1 Problems of Definition

The *World Book Encyclopedia* reads: "adaptation, in biology, is the process by which a living organism becomes better fitted to survive in its environment (surroundings)." Also, according to the glossary edited by the International Union of Physiological Sciences (IUPS; see Bligh and Johnson 1973) adaptation is "a change which reduces the physiological strain produced by a stressful component of the total environment." None of these definitions uses the term *adaptation* in the precise Darwinian sense of genetically fixed changes transmitted to the offspring. Further, the IUPS Glossary makes clear that "there are no distinct terms which relate to genotypic adaptations .... All such genetically fixed attributes of a species or subspecies are covered by the general term *genotypic adaptation,*" whereas "*phenotypic adaptations* refer to changes occurring within the lifetime of the organism." Hence, in this book, no explicit Darwinian connotation will be implicated in the word adaptation.

According to the above definitions, the fundamental basis for assessing adaptation involves the evaluation of a benefit relative to the environmental stress. In contrast, the definition proposed by Adolph (1956) discards the notion of benefit: "adaptations are modifications of organisms that occur in the presence of particular environments or circumstances ... not limited, as is often done, to modifications that seem favorable to the individual." The present book conforms to this last definition. In fact, descriptions of responses to high altitude "have often omitted explicit consideration of relative benefit, or ascriptions of adaptive value have been based on unstated or unsupported assumptions with regard to selective advantage or enhanced performance" (Mazess 1975).

It must be clear, therefore, that the adaptations to be considered throughout this book are functional adjustments that do not imply in an obligatory sense that they are beneficial.

1.2.2 Problems in Assessing Adaptation to High Altitude

As with environmental adaptation in general, the assessment of adaptation to high altitude involves several specific problems and difficulties. These can be outlined, and the scope of this book narrowed, if we pay some attention to the following questions: What adapts to what? How? In what course of time?

1.2.2.1 What Adapts to What?

It is common knowledge that vertebrates (and most invertebrates) depend on the availability of oxygen for oxidative metabolism at the cellular utilization sites. On the other hand, as previously stressed (Sect.1.1.2), decreased availability of ambient O_2, along with reduced barometric pressure, is the common denominator common to high-altitude habitats. Since adaptagents are specific changes in environmental factors (Adolph 1964), and since animal organisms take up oxygen from the ambient milieu, the partial pressure of O_2 can be viewed as the major adaptagent to be dealt with: "Oxygen tension is everything; barometric pressure in itself does nothing or almost nothing" (Bert 1978, p 540). Accordingly, inspired P_{O_2} is shown on the top of Fig. 1.5, which is a flow diagram of the path that oxygen takes from the ambient medium to the cell machinery.

Adaptations to low P_{O_2} can be differentiated into two broad categories. First are the changes in the mechanisms concerned in O_2 transport from inspired medium to cells. These form the bulk of Chaps. 3–5; most of them are shown in Fig. 1.5. In the second category are the changes in the biochemical processes of O_2 utilization at the intracellular level, in mitochondria and other loci; these will be considered in Chap. 6. In either category the results of adaptations can be labeled *adaptates,* a term proposed by Adolph (1956) to designate "any measurable modification in structure of function when adaptagent impinges."

Adaptates vary as a function of the stimulating value of the adaptagent involved (hypoxia in this book), depending on the intensity, duration, frequency, and variability of that adaptagent (Adolph 1964) and on the nature of the possible interactions at work (such as hypoxia and cold or exercise, hypoxia and malnutrition, etc.). This sometimes makes it difficult to summarize the data obtained under altitude hypoxia, for there are as many protocols as experimenters.

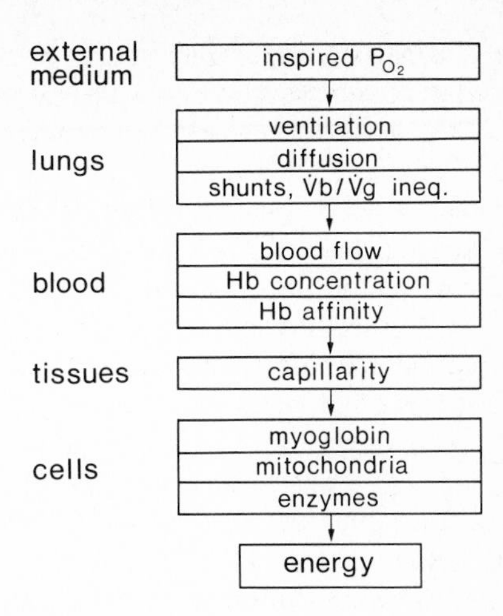

Fig. 1.5. Schematic representation showing from above downward a flow diagram of the path oxygen takes from external medium to cell enzymes for energy production. (Derived from Monge and Whittembury 1976)

For a study of the physiological effects of low P_{O_2} alone, high altitude simulation in decompression chambers is thought to be the "experimental ideal" (Monge and Whittembury 1976). Yet not only P_{O_2} but also transient changes in the total pressure, P_B, may affect the experimental subject (Grover et al. 1982). Moreover, the operation of pumps to depressurize and ventilate the chambers may generate noise and vibrations which, in association with uncontrolled changes in ambient temperature, constitute superimposed stressors. Then the question arises as to whether some reported adaptations are due to the interaction of concurrent adaptagents, or relate to what is confusingly called the "stress syndrome," rather than to the specific effects of hypobaric hypoxia. Only carefully planned experiments can resolve these probelms.

1.2.2.2 How Do Adaptations Occur?

The fundamental approach to this question and the associated concepts can be found in Adolph (1964). Some terms to be used in following chapters can be introduced by the following very simple model

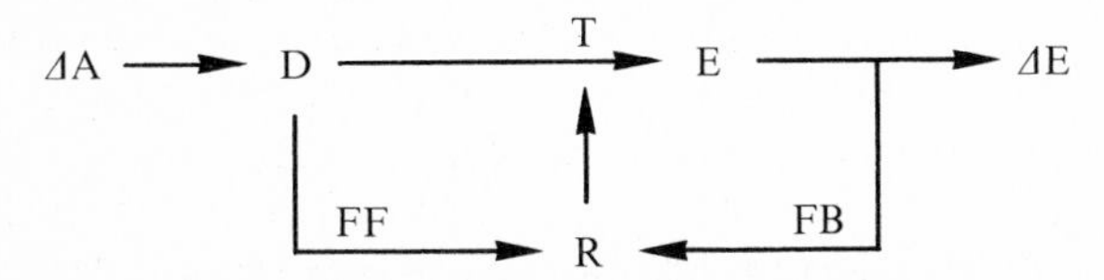

where ΔA is an adaptagent that relates to changes in a specific environmental factor, and is sensed by some detector D within the organism. Information then passes to an effector E, via a transmitting system T over which control is exerted by some regulating mechanisms R. The consequence is that activity of E changes and, therefore, adaptate ΔE is elicited. Transmitter T may involve many steps, not always properly identified. Regulator R is off the line of communication from D to E, and receives information either from the output of E (feedback mechanism, FB) or from D (feedforward mechanism, FF) or from both.

Although one may accept this very simple model for the purpose of definitions, its translation into physiological terms is not always clear. Theoretically, there may be as many varieties of the adaptational elements as there are adaptagents and adaptates. As a consequence of the large number of detector and effector units, there is a multiplicity of signal pathways, with possible overlaps among them. For instance, as will be seen later, some adaptational responses to hypoxia in natural mountainous environments are not wholly separate from adaptations to cold. On the other hand, nonspecific responses such as the so-called general adaptation syndrome, may be part of adaptive events in the earlier stages of exposure to high altitude. They occur preferably when severe stress is imposed on the organism, and they are an expression merely of damage, not of defense. The specificity of adaptive response to an adaptagent also varies with its quality and strength, and gradually improves with increasing duration of the exposure, provided that the adaptagent does not exceed a critical level.

Most of this book deals with ΔA as the decrease in ambient P_{O_2} when ascending to high elevations within the altitude range where man and other vertebrate animals are normally living. Some experiments, however, will be reported during which the hypoxic stress was exceedingly severe in comparison with natural situations. When properly identified, interacting factors, such as hypoxia, cold, dehydration, caloric restriction, deserve consideration. On the other hand, the E and ΔE, which will be dealt with, are essentially the structural and functional components interposed on the path taken by oxygen from external medium to the intracellular sites of utilization, some of which are shown in Fig. 1.5. Many adaptates, therefore, are expected to be elicited in responses to hypoxia. Correspondingly, adaptation will not be regarded as a single process.

1.2.2.3 Time Courses of Adaptations

Functional adaptive changes require time to materialize. They can occur through the following processes.

a) Accommodations, also called adjustments, are the initial responses to acute change in the environment, as experienced, for instance, by the lowlander rapidly translocated to high altitude, or by the experimental animal in a recently depressurized low-pressure chamber. Such initial responses occur within time intervals ranging from seconds to minutes, sometimes hours, depending on the intensity of the adaptagent (mainly hypoxia) and the adaptate under consideration.

b) Acclimations and acclimatizations are adaptational responses that occur over a period of days to weeks or months. According to the IUPS glossary (see Sect. 1.2.1), acclimations are physiological changes caused in the laboratory by *experimentally induced* variations in a single environmental factor, while acclimatizations are physiological changes caused by *natural* variations in the entire natural environment. Like accommodations, these processes develop within the lifetime of an organism, and may reverse rapidly when the adaptagent is returned to normal (here normoxia). They all are *phenotypically labile adaptations,* distinct from fixed genotypic adaptations.

c) Genotypic adaptations result from genetically fixed attributes in those species that have lived for generations in their environment. Here the term adaptation is used in a precise Darwinian sense. In the following the term will refer to structural, physiological and/or biochemical features that are inherited. These features apparently allow some individuals to cope more successfully than other with the adversities of chronic deprivation of oxygen. They do not reverse with descent to sea level.

1.3 A Comparative Account of Respiratory Processes

A comprehensive review of the accumulated literature concerned with adaptation to high altitude, although presenting an interesting and formidable challenge, is clearly beyond the scope of this monograph. Particular attention will be given here to respiratory processes. Since "the diminution of barometric pressure acts upon living beings only by lowering the oxygen tension in the air they breathe, and in the blood which supplies their tissues" (Bert 1978; p 1036), emphasis will be placed on the adaptations of the convection and diffusion mechanisms of the oxygen transfer from ambient medium to cells. The adaptations to changing availability of ambient O_2 will be considered over periods of minutes, hours, weeks, years, and generations within a species, and among vertebrates. As far as possible, a comparative approach will be used, as it is true that: (1) extrapolation from one group of organisms to another can lead to biological generalizations that cannot be reached by other methods, (2) different kind of organisms solve similar life problems in different ways (Prosser 1964). Emphasizing differences as well as similarities among organisms, the comparative approach is not based, as sometimes argued, on the theory that only a single optimum adaptation exists for a given adaptagent.

Chapter 2

The Respiratory Gas Exchange System and Energy Metabolism Under Altitude Hypoxia

As pointed out in Chap. 1.1.1, high altitude is inescapably associated with lowered pressure of oxygen. This makes oxygen transfer into and utilization by the tissues more difficult. The ensuing state of hypoxia, as here examined (Sect. 2.3), becomes a factor of great concern when oxygen demand increases during exercise or in cold-induced thermogenesis. Then, unless the inhaled medium is artificially supplemented with oxygen, organisms rely solely upon the adaptation of the mechanisms of the O_2 transfer among the various compartments of the respiratory gas exchange system. These are presented in the next two sections, which are adressed to those biologists who are not familiar with the principles of comparative respiratory physiology, concentrating on some general aspects in respect of which altitude hypoxia differs from sea level normoxia.

2.1 Concept of Conductance

2.1.1 From External Medium to Mitochondria

As extensively discussed elsewhere (e.g., Dejours 1981), the modern animal physiologist views the living body as a system of oxygen (O_2) and carbon dioxide (CO_2) transfer. The O_2 and CO_2 molecules must be transferred by *diffusion* and by *convection* through various structures and compartments between the cells and the ambient medium. These are shown for a mammal in Fig. 2.1.

In other vertebrates, either lung breathers or gill/skin breathers, the general view outlined in Fig. 2.1 is valid. The external gas-exchange organs, however, are different, which has implications with regard to gas-exchange efficacy (Sect. 2.2.1).

The embryo exchanges O_2 and CO_2 essentially by diffusion. This process takes place through the shell, shell membranes, and chorioallantoic membranes of bird and reptile eggs (Rahn et al. 1979). Early in the incubation period, however, the beating heart starts to circulate blood through the capillaries of the embryo, thus adding a convective respiratory process. Similar processes occur for fetal and uterine O_2 and CO_2 exchange through the placenta: diffusion takes place from the maternal to the fetal side of the placental exchanging surface; fetal and maternal circulation ensure O_2 and CO_2 convective transport by blood (see Bartels 1970).

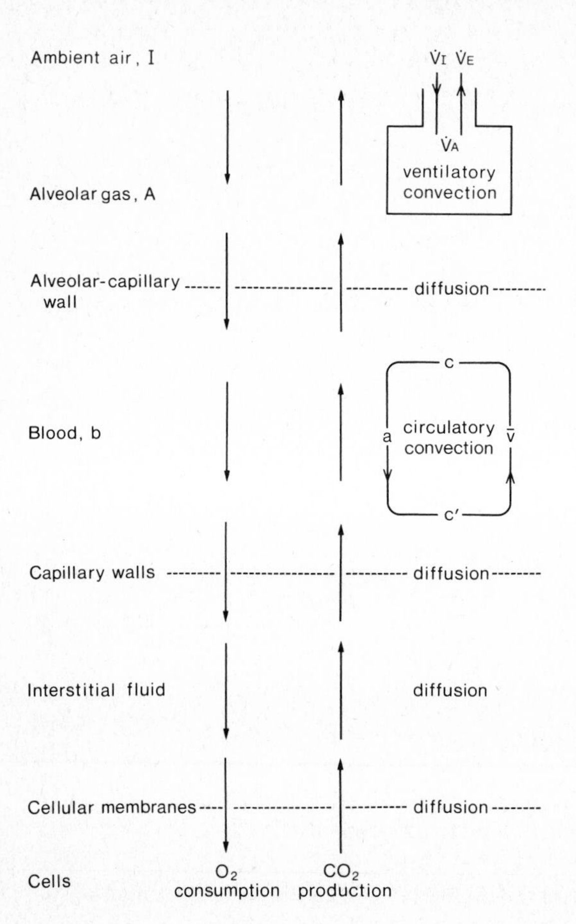

Fig. 2.1. Scheme of the O_2 and CO_2 transfer system of mammals. (Dejours 1981)

2.1.2 Basic Equation

For the description of gas transfer in steady state, a basic equation discussed by Piiper et al. (1971) and Dejours (1981) is the following

$$\dot{M}_x = G_x \cdot \Delta P_x \qquad (2.1)$$

or

$$\Delta P_x = \frac{1}{G_x} \dot{M}_x , \qquad (2.1b)$$

in which $\dot{M}_x$ is the amount of substance x (O_2 or CO_2) transferred per unit time (=rate of transfer; mmol · min^{-1}) between two locations in the gas exchanges system, G_x ($mmol \cdot min^{-1} \cdot Torr^{-1}$) is the conductance of the substance x and ΔP_x (Torr) the difference of partial pressure between the two locations.

The conductance, therefore, is the amount of the substance x (O_2 or CO_2) transferred from one location to the other per unit time under a unit driving partial pressure

$$G_x = \frac{\dot{M}_x}{\Delta P_x}. \tag{2.2}$$

The reciprocal of the conductance, $1/G_x$ in Eq. (2.1 b), is analogous to a resistance, with ΔP_x being analogous to an electrical potential difference, and $\dot{M}_x$ to an intensity. The lower this resistance (the greater the conductance), the smaller the required ΔP_x value for a given value of $\dot{M}_x$.

2.1.3 Diffusive Conductances

Diffusive conductances, G_x in Eq. (2.2) (sometimes written $Gdiff_x$), depend upon three factors: (1) the geometry of the structures through which diffusion takes place, A/E, with A being the surface area and E the thickness, (2) the diffusion coefficient, D_x, as defined by physicists, (3) the membrane capacitance coefficient, β_x, as defined by Eq. (1.3)

$$Gdiff_x = \frac{A}{E} \cdot D_x \cdot \beta_x. \tag{2.3}$$

In a gas phase, the diffusion coefficient, D_x, is inversely proportional to the barometric pressure (Reid and Sherwood 1966). This is important in the gas exchange across eggshells (Rahn 1977, Sect. 5.2.1).

In all those biological structures that oppose resistance to diffusion of respiratory gases, the exact values of the diffusion and capacitance coefficients are most often unknown. Some values of their product, $D_x \cdot \beta_x = K_x$, the Krogh's diffusion constant, have been determined for the in vitro diffusion of O_2 or CO_2 through thin layers of various tissues (Kawashiro et al. 1975).

Most often, none of the factors involved in Eq. (2.3) is known. Then, the diffusion conductance is determined according to Eq. (2.2) as the ratio $\dot{M}_x/\Delta P_x$.

2.1.4 Convective Conductances

For the long-distance transport from the ambient medium to the external gas exchange surfaces, on the one hand, and from these surfaces to the systemic capillaries, on the other hand, vertebrates rely on convective ventilatory flow and blood flow.

The governing (Fick's) equation for convection in steady state can be written

$$\dot{M}_x = \dot{V} \cdot (Cin_x - Cout_x), \tag{2.4}$$

where $\dot{V}$ is the flow rate and Cin_x and $Cout_x$ are the concentrations of the species x (O_2 or CO_2) in the medium that enters and leaves the gas exchange area. Equation (2.4) is not changed if rewritten as follows

$$\dot{M}_x = \dot{V} \frac{Cin_x - Cout_x}{Pin_x - Pout_x} (Pin_x - Pout_x) \tag{2.5}$$

or

$$\dot{M}_x = \dot{V} \cdot \beta_x \cdot (Pin_x - Pout_x) \tag{2.5b}$$

in which $Pin_x - Pout_x$ is the partial pressure difference of x between the inflowing and outflowing medium, and β_x is the capacitance coefficient of that medium for x [Eq. (1.3)]. Comparison of Eqs. (2.1) and (2.5 b) shows that

$$G_x = \dot{V} \cdot \beta_x . \tag{2.6}$$

Therefore, convective conductances, G_x (sometimes written $Gconv_x$) depend upon both the flow rate, $\dot{V}$, and the capacitance coefficient of the medium, β_x.

In air (as in any gas phase) and in water, the O_2 and CO_2 capacitance (solubility) coefficients are unaffected at high altitude: at steady temperature, they remain constant whatever the elevation is. Consequently, the ventilatory O_2 conductance ($Gvent_{O_2}$) changes only when the ventilatory flow of the external medium [$\dot{V}$ in Eq. (2.6)] changes.

In blood, the O_2 and CO_2 capacitance coefficients describe not only physical solubilities, but also chemical binding with hemoglobin. Their values change with altitude as a function of many factors (Chap. 4), and so does the value of β_x in Eq. (2.6). Consequently, unlike $Gvent_{O_2}$, the circulatory O_2 conductance (named perfusive conductance, $Gperf_{O_2}$) can change either through the combined changes in both the circulatory flow and blood O_2 capacitance, or through the independent change in any of these two factors.

2.1.5 Adaptation to Hypoxia

Since the reciprocals of conductances, i.e., resistances (Sect. 2.1.2), are additive when in series as in the respiratory gas exchange system (Fig. 2.1), any increase in either one of the above-mentioned O_2 conductances will lower the resistance that opposes to the O_2 flux ($\dot{M}_{O_2}$) from external medium to cells, thereby favoring adaptation to hypoxia. Evidence for the existence of such an adaptation at high altitude is provided in the general approach that follows.

2.2 The O_2 and CO_2 Cascades

Figure 2.2 gives an example of the O_2 and CO_2 partial pressures in the various compartments of the gas-exchange system. This example pertains to a resting mammal studied in steady state first near sea level (P_B = 750 Torr; solid lines),

Fig. 2.2. Partial pressures of O_2 and CO_2, P_{O_2}, and P_{CO_2}, in inhaled air and in the various compartments of the gas-exchange system of an awake, resting dog, studied first near sea level (PB = 750 Torr, *solid lines*), then after 2 weeks at the simulated high altitude of 4.5 km in a hypobaric chamber (PB = 433 Torr, *dashed lines*). The partial pressures of O_2 and CO_2 in the interstitial fluid and cells are hypothetical; in other compartments they are personal, unpublished data

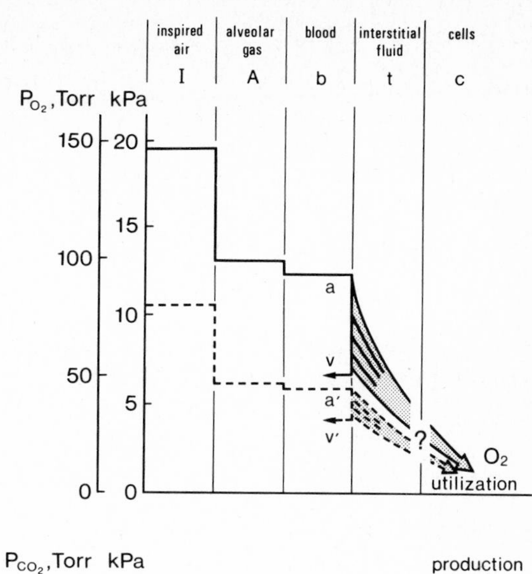

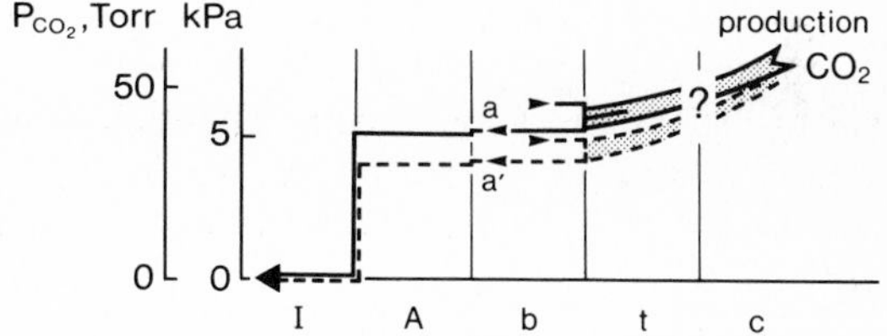

then after two weeks in a hypobaric chamber operated at 4.5 km simulated altitude (PB = 433 Torr; dashed lines). For the interstitial and cellular compartments, the values are hypothetical.

Focusing attention on the solid lines in the upper part of the figure, it is clear that several steps occur in the transfer of oxygen from the environment to cells. Each step in the O_2 partial pressure (ΔP_{O_2}) from one compartment to the next is proportional to the resistance opposed to the O_2-flux.

Comparing dashed and solid lines in the upper part of Fig. 2.2, it is also clear that the O_2 pressure head in the inspired air drops with ascent to high altitude. Furthermore, the various ΔP_{O_2} steps from one compartment to the other are reduced. This suggests that the resistances opposed to the O_2-flux are less at high altitude than at sea level, in so far as the O_2 consumption, $\dot{M}_{O_2}$, remains unaffected. In other words, the various O_2 conductances appear to be greater at high altitude, as are the O_2-conductances per unit of oxygen consumed, i.e., the $\dot{M}_{O_2}$-specific conductances ($G_{O_2}/\dot{M}_{O_2}$ in Table 2.1). These changes are of adaptive significance. Underlying mechanisms will be examined in subsequent chapters, as will those responsible for the accompanying hypocapnia (reduced CO_2 partial pressures) observed at high altitude (Fig. 2.2, lower part; compare dashed with solid lines). Some details, salient features among vertebrates, and concepts to be used later are introduced below.

Table 2.1. Transport of oxygen between air and the tissues of an awake dog resting in a hypobaric chamber operated first near sea level (s.l.; P_B = 750 Torr), then at 4.5 km simulated high altitude for two weeks (h.a.; P_B = 433 Torr; personal, unpublished data from Fig. 2.2). Pin_{O_2} and $Pout_{O_2}$ refer to the partial pressures of oxygen at the various locations in the gas exchange system specified by bracketed symbols; $\Delta P_{O_2} = Pin_{O_2} - Pout_{O_2}$ is proportional to the resistance opposed to the O_2-flux; $G_{O_2}/\dot{M}_{O_2} = 1/\Delta P_{O_2}$ [according to Eq. (2.1)] is the $\dot{M}_{O_2}$-specific conductance between the two locations under consideration. Note that ΔP_{O_2} decreases, and therefore $G_{O_2}/\dot{M}_{O_2}$ increases at high altitude (see text for further comments)

		Pin_{O_2}	$Pout_{O_2}$	ΔP_{O_2}	$G_{O_2}/\dot{M}_{O_2}$
		Torr			$Torr^{-1}$
I, A	convective transport	(I)	(A)		
	s.l.	147	98	49	0.0204
	h.a.	80	47	33	0.0303
	h.a./s.l.	0.54		0.65	1.49
A, a	complex transport	(A)	(a)		
	s.l.	98	94	4	0.2500
	h.a.	47	44	3	0.3333
	h.a./s.l.			0.75	1.33
a, $\bar{v}$	convective transport	(a)	($\bar{v}$)		
	s.l.	94	50	44	0.0227
	h.a.	44	31	13	0.0769
	h.a./s.l.			0.30	3.39
I, $\bar{v}$	overall transport	(I)	($\bar{v}$)		
	s.l.	147	50	97	0.0101
	h.a.	80	31	49	0.0204
	h.a./s.l.			0.51	2.02

2.2.1 O_2 Loss in the External Gas Exchange Organ

Focusing on the first ΔP_{O_2} step in the gas exchange system, between the inspired air (I) and the alveolar gas (A), it appears from Fig. 2.2 that approximately one-third of the inhaled O_2 molecules were lost before they reached the dog's pulmonary gas-exchanging surfaces [the O_2 concentration is proportional to P_{O_2}; Eq. (1.3)]. The functional consequence is that the pressure head for subsequent O_2 transfer into the organism was greatly decreased (Fig. 2.2, compare A with I). Do other vertebrates suffer from a similar handicap?

To answer this question, one may consider a few simplified models of gas-exchange organs deduced from their complicated anatomical structure (Fig. 2.3). In these models, flowing internal medium (blood, b) comes into diffusional exchange contact (D) with an outer medium (gas, g, or water, w) also streaming in most cases.

a) Alveolar lungs are found in mammals (Fig. 2.3, left), reptiles, and amphibians. They exist also in some fishes; but these do not live at high altitude. As a typical example, the mammalian lung can be viewed as a chamber within the body open to the outside by a single channel, the trachea. The conducting airways

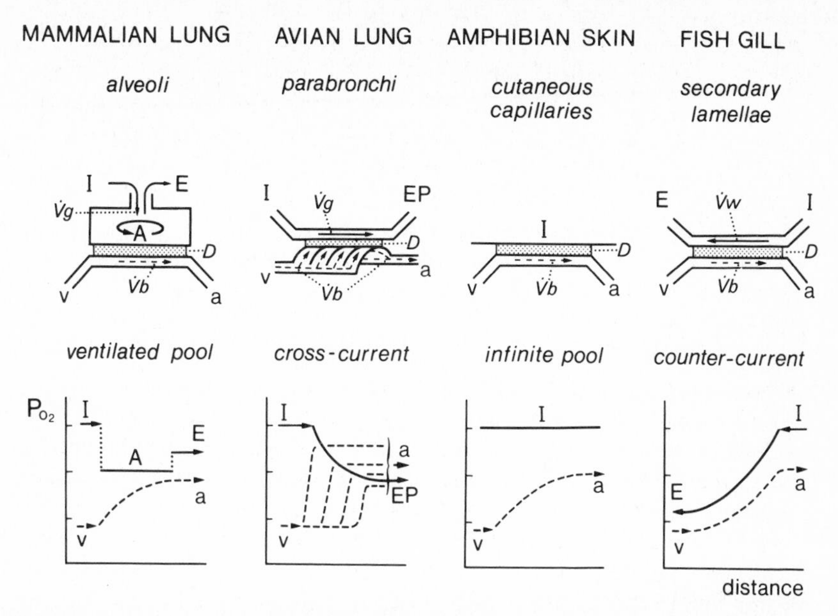

Fig. 2.3. The four major models of external gas exchange among vertebrates: *from left to right* ventilated pool, cross-current, infinite pool, counter-current. (Inspired by Piiper and Scheid 1977). *Top* scheme of the contact area between blood and external medium, air in the mammal and bird, water in the amphibian and fish. *Bottom* schematic curves showing partial pressure changes for O_2 in external medium and blood in the gas-exchange area. *I* incoming medium; *E* expired medium; *A* alveolar gas; *EP* end-parabronchial gas; *v* and *a* afferent (venous) and efferent (arterial) blood, respectively; $\dot{V}g$, $\dot{V}w$, and $\dot{V}b$ flow rates of gas, water and blood respectively; *D* diffusive barrier. See text for further explanatations

which follow the trachea form a more or less branching system by irregular dichotomy. They contain a volume of gas, the dead space, that does not participate in the O_2 and CO_2 molecules exchanges. Gas exchange takes place only in the terminales bronchioles, alveolar ducts, and alveolar sacs surrounded by blood capillaries. A to-and-fro mode of ventilation (tidal ventilation) is therefore necessary. The expiratory phase is functionally equivalent to a breath-holding period, and the same is true for the early part of inspiration due to reinhalation of the gas of the dead space, gas vented from the alveoli during preceding expiration. During these periods, the O_2 partial pressure in the alveolar gas ($P_{A_{O_2}}$) decreases, and that of CO_2 ($P_{A_{CO_2}}$) increases, because O_2 is continually being removed by the blood and CO_2 added. It is only when the gas of the dead space has been washed out that fresh air replenishes the oxygen and removes CO_2. Therefore, the O_2 partial pressure in alveolar gas, $P_{A_{O_2}}$, results from the balance between the rate at which O_2 is removed from gas to blood, and the rate at which it is replenished from the ambient air through tidal ventilation: the greater the ventilation, the greater $P_{A_{O_2}}$ and, consequently, the lower the P_{O_2} difference from inspired air to alveolar gas. In the animal shown in Fig. 2.2 (upper part), the balance was such that $P_{A_{O_2}}$ was lower than $P_{I_{O_2}}$ by 49 Torr near sea level, but by only 33 Torr at high altitude (Table 2.1). There was, therefore, a hyperventilatory response during the hypobaric experiment. Due to that hyperventilation, CO_2 was removed in greater amounts, with ensuing hypocapnia (Fig. 2.2, lower part).

Figure 2.3 (lower row, far left) shows schematically the variations of blood O_2 pressure from venous (v) to arterial (a) blood along the pulmonary capillary. Note that arterial P_{O_2} is lower than alveolar P_{O_2}.

b) Tubular lungs are found in birds (Fig. 2.3, second column). Ventilation is tidal, as in mammals, but the functional "breath-holding" period encountered in alveolar lungs during expiration is virtually absent, due to the unidirectional ventilation of the parabronchial gas-exchanging surfaces, which proceeds from caudal to cephalad during expiration as well as during inspiration (review by Scheid 1979). Furthermore, the blood/air barrier is thinner in birds than in mammals, and the ratio of exchange surface area/lung volume is very high (Dubach 1981; Duncker 1972). On the other hand, in the avian lung, ventilatory gas flow ($\dot{V}g$) and blood flow ($\dot{V}b$) cross each other in the gas exchange region (*cross-current system;* see Scheid 1979). Figure 2.3 shows schematically the anatomical arrangement and the resulting P_{O_2} profile in the gas phase from the initial (I) to the end-parabronchial value (EP). Note the cross-over of the O_2 partial pressures in parabronchial gas (lower raw, solid line) and blood (dashed lines). There is no abrupt step in the gas P_{O_2} values; instead, a progressive decline is evident. Note also that arterial blood is a mixture from all the blood capillaries in contact with the various segments of the gas-exchanging region. The consequence is that P_{O_2} in arterial blood may exceed that in the exiting parabronchial gas. The possibility exists that P_{O_2} in the arterial blood is greater in birds than in mammals at identical elevation, and therefore at identical inspired P_{O_2} (compare the levels of a in the two far left columns of Fig. 2.3).

c) Skin breathing is important for exchange of O_2 and CO_2 in most amphibians and some fishes. Gas exchange takes place between the cutaneous, subepidermal, capillary network and the ambient air or water. Although there is no ventilation in the sense of ventilatory flow produced by respiratory movements, the state is functionally equivalent to infinite ventilation (Piiper and Scheid 1977). The consequence is that the pressure head for O_2 diffusion is the partial pressure of O_2 in the ambient medium (I in Fig. 2.3, third column). This might be of a great advantage since no oxygen is lost, as in the external gas organ of mammals. The skin, however, constitutes a much thicker diffusion barrier than the gas/blood barrier in alveolar or tubular lungs. As suggested in Fig. 2.3, therefore, the O_2 transfer is diffusion-limited, and P_{O_2} in the arterial blood is far below that in the ambient medium.

d) Gills are efficient gas exchangers in water breathers (Fig. 2.3, last column). In fish, the anatomical arrangement of the gill apparatus and the operation of a double-pumping bucco-opercular mechanism are such that (1) water flows unidirectionally across the gills from the buccal to the opercular cavities through the ventilatory cycle; (2) flows of water (w) and blood (b) are in opposite directions (*counter-current system;* see Hughes and Morgan 1973). The functional result is that the blood leaving a gill capillary may be brought very close to fresh water entering the gills. Theoretically, therefore, the value of P_{O_2} in the arterial blood, a mixture from all blood capillaries, could be very close to that in the inspired water, a potentially decisive advantage. However, the existence of diffusion barriers, regional inhomogeneities and functional blood shunts (Sect. 2.2.2) may lead to a reduction in gas exchange efficiency.

e) Comparative analysis of efficiency of gas exchange in vertebrates' gas-exchange organs has been made in the general terms of conductances, relative partial pressure differences and limitations attributable to ventilation, to medium/blood transfer and to perfusion by Piiper and Scheid (1971, 1975, 1977, 1981). Based on some of the simplified models shown in Fig. 2.3, this analysis indicates that the external gas exchange systems of vertebrates can be arranged into the following order of decreasing efficiency: counter-current > cross-current > ventilated pool. However, when applied to the available physiological data from a few species, the analysis shows that the differences in efficiency actually attained are much less pronounced. They depend to a large degree on the values of ventilation, perfusion, and diffusion conductances, and they may be even less marked when taking into account functional inhomogeneities. Some of these are considered below.

2.2.2 Complex Gas/Blood Transfer

Returning to the mammalian lung, there is a second, smaller step in the P_{O_2} cascade shown in Fig. 2.2, between the alveolar gas (A) and arterial blood (a; see also Table 2.1). As for CO_2 in the reverse direction, the arterio-alveolar difference is at the limit of measurability (Fig. 2.2, bottom). These unequal P_{O_2} and P_{CO_2} differences may arise by means of several mechanisms, a short outline of which is given below (for complete analysis see Rahn 1949; Rahn and Farhi 1964; West 1977).

a) Diffusion limitation is a first mechanism to be considered. The diffusion processes of the O_2 and CO_2 from gas to blood, and vice versa, come up against many serial resistances opposed by the various structures collectively designated as the gas/blood barrier. These structures are essentially made of aqueous media which, at the body temperature, are about 30 times less permeable to O_2 than to CO_2 under the same unit partial pressure gradient (see Dejours 1981). This is one reason for the (A-a)P_{O_2} difference being larger than the (a-A)P_{CO_2} difference (Fig. 2.2). The former may be particularly important at high altitude when the diffusive conductance (also called lung-diffusing capacity) is low. For instance, in heavy exercise at high altitude, when the amount of O_2 that must be transferred increases considerably, while the time contact between blood and capillary wall drops due to increased cardiac output, the O_2 pressure at the end of the pulmonary capillaries may remain far below the alveolar O_2 pressure (see Fig. 5.4 D).

b) Nonhomogeneity in the distribution of alveolar ventilation ($\dot{V}_A$) *and blood perfusion* ($\dot{V}b$) is a second mechanism affecting the alveolar-arterial P_{O_2} and P_{CO_2} differences. Among other things, the alveolar-arterial transfers depend on the rate of renewal of each medium, gas and blood. In practice, even in the "normal" lung, the share of the total blood flow going to each unit (alveolus) is different, as is also its share of the total ventilation, i.e., individual $\dot{V}_A/\dot{V}b$ ratios are far from being identical in all parts of the lungs. It has been shown that these regional inequalities of the $\dot{V}_A/\dot{V}b$ ratio (distribution factor) affect the O_2 transfer more than that of CO_2 transfer; the P_{O_2} of the mixture of the various bloods differs by a few Torr from the P_{O_2} of the mixture of the various alveolar gases, whereas the (a-A)P_{CO_2} difference is very small. The generally accepted view is that the distribution factor is of little importance at high altitude, although some findings conflict with this.

c) Venous admixture, or the addition of mixed venous blood to arterial blood, is a third mechanism affecting alveolar-arterial differences. Such a venous contamination (shunt) may occur through true anatomical shunts between the pulmonary artery and veins, and through venous drainage of bronchial or myocardial blood. Due to the shape of the oxygen dissociation curve of blood (C_{O_2} vs. P_{O_2} relationship), however, the alveolar-arterial P_{O_2} difference resulting from venous admixture is usually small in hypoxia.

In nonmammalian vertebrates, either air-breathers or water-breathers, the above three mechanisms are believed to function similarly in determining the P_{O_2} and P_{CO_2} differences between the internal (blood) and external (air or water) fluids in diffusional contact. Present knowledge, however, is extremely limited (birds: Powell and Wagner 1982; reptiles: see Wood and Lenfant 1976; fishes: see Randall 1970).

2.2.3 O_2 Transfer into the Tissues

Figure 2.2 (upper part) makes clear that tissue gas exchange begins at the arterial inlet to the capillary bed, P_{O_2} falling rapidly along the systemic capillary because oxygen moves out into the interstitial fluid and cells. Consequently, a P_{O_2} step occurs from arterial blood (a) to venous blood (v). Note that the P_{O_2} step is smaller at high altitude (dashed vertical line) than near sea level (solid vertical line). This indicates that the $\dot{M}_{O_2}$-specific conductance is greater ($Ga,v_{O_2}/\dot{M}_{O_2}$ in Table 2.1). Underlying mechanisms will be considered in Chaps. 4 and 5.

Although hypothetical, the P_{O_2} profiles in tissues shown in Fig. 2.2 (t and c compartments) conform with present knowledge (Grunewald and Sowa 1977; Lübbers 1977). They illustrate the fact that O_2 moves out from the systemic capillary into the cells, by physical diffusion from an area of high O_2 pressure to one of low pressure. Note that the P_{O_2} profile for tissues is shifted downward at high altitude. Its lower boundary, however, is not as low as it would be, were the arterio-venous P_{O_2} step not reduced as it is.

2.2.4 Elimination of Carbon Dioxide, and Acid Base Status

Figure 2.2 shows that (1) the slopes of the P_{CO_2} profiles between the intracellular sites of CO_2 production and capillary blood are not as steep as those of the P_{O_2} profiles, and (2) the (v-a)P_{CO_2} differences are smaller than the (a-v)P_{O_2} differences. The reason is that the capacitance of the various body fluids is greater for CO_2 than for O_2.

Also shown in Fig. 2.2 (lower part) is hypocapnia at high altitude, i.e., lowered values of P_{CO_2} in blood and alveolar gas due to the hyperventilatory response to hypoxia. Therefore, the acid-base status is modified, since the pH does increase at first when P_{CO_2} decreases, according to the Henderson Hasselbalch equation. How far toward alkalosis the pH changes at high altitude, due to a given drop of P_{CO_2}, depends upon the concomitant change in the concentration of bicarbonate, and therefore upon the buffer value as defined by the slope of the $[HCO_3^-]$ vs. pH relationship.

Over recent years, renewed attention has been paid to the acid-base balance in animals with variable temperature (Reeves 1977; White and Somero 1982). Of special interest here are those ectothermic animals living, for instance, at 4.3 km altitude, on the altiplano of Peru. When they emerge from a ground burrow to bask in direct sunlight, their body temperature may go from 5° to 30 °C in approximately half an hour (Pearson and Bradford 1976). According to Reeves (1977), such an increase in body temperature is associated with a dramatic change in the acid-base status of blood (pH decreasing from 7.93 to 7.47, and P_{CO_2} increasing from 5.8 to 23.7 Torr), but the fractional dissociation of histidine imidazole groups of proteins remains constant, hence also the mean protein net charge; this would defend neutrality, or near neutrality, which is thought to represent the ideal intracellular pH for optimal enzyme reactions.

2.3 Energy Expenditure and O_2 Availability

Although unexplained energy appears to be produced during muscular contraction under a variety of conditions (Curtin and Woledge 1978), the hydrolysis of ATP (adenosine triphosphate) is classically viewed as the single source of the free energy transduced in the fundamental energy-requiring processes. In a useful approximation, therefore, the overall energy output per unit of time ($\dot{E}$) can be visualized with the aid of the following equation

$$\dot{E} \propto \dot{M}_{ATP} = p\dot{M}_{AN} + q\dot{M}_{O_2}, \tag{2.7}$$

in which $\dot{M}_{ATP}$, $\dot{M}_{AN}$, and $\dot{M}_{O_2}$ are the rates of ATP splitting, anaerobic, and oxidative (aerobic) ATP-yielding reactions, respectively, and p and q coefficients are the amount of ATP resynthetized per unit of anaerobically formed products, or per unit of O_2 consumed.

Since energy metabolism is an appropriate indicator of the physiological state of an organism, the effects of acute and chronic hypoxia on both the aerobic and anaerobic components of energy expenditure are examined next.

2.3.1 Aerobic Metabolism

It is common knowledge that physiological "regulators" are distinguished from "conformers" by their ability to compensate for change (within certain limits) of environmental factors. As for change in the ambient O_2 availability, some fish are oxygen-dependent, i.e., conformers, but all aerial vertebrates are thought to be oxygen-independent, i.e., regulators (see Prosser 1973). In these latter, however, the metabolic uptake of oxygen [$\dot{M}_{O_2}$ in Eq. (2.7)] may be limited by sufficiently hypoxic environments. In other words, in all aerobic organisms there exists a critical ambient oxygen pressure below which O_2 consumption decreases, thus suggesting that the O_2 conductance of the gas-exchange system [G_x in Eq. (2.1)] becomes limiting. This critical pressure might provide one measure of regulation or resistance to hypoxia. Yet it is difficult to evaluate precisely, for its value may vary with a number of factors. As a general rule, however, the higher the O_2 needs of the body, the higher the critical O_2 pressure. The following two sections illustrate this point.

2.3.1.1 Minimal Oxygen Consumption

The standard $\dot{M}_{O_2}$, i.e., the minimal oxygen uptake measured at rest, in post-absorptive conditions and at neutral ambient temperature (Kleiber 1961), appears to be remarkably constant in mammals, including man, over the wide range of ambient P_{O_2} encountered in mountains (Stickney and van Liere 1953; Monge and Whittenbury 1976). The same may be true for birds (Black and Tenney 1980) and reptiles (Bennett and Dawson 1976; Jameson et al. 1977) but data are scarce.

As a typical example, Fig. 2.4 shows that even small mammals can sustain a considerable reduction in ambient P_{O_2} with no reduction in the standard $\dot{M}_{O_2}$, when, and only when, studied at thermoneutral temperature: in the rats studied by Arieli et al. (1977) at 30 °C, there was no evidence for the existence of a critical partial pressure of O_2 down to 32 Torr, i.e., a simulated altitude above 10 km. Whether adaptation to high altitude results in an improved capacity for maintaining aerobic metabolism down to a lowered critical P_{O_2} is a fundamental question, as yet unanswered. Near basal conditions, highland rodents have been reported to be more resistant to hypoxia than lowland species (Morrison 1964; Hall 1966).

Comparison of two groups of organisms, one from sea level and the other from high altitude, is complicated by the fact that their body masses are different, since growth and development are impaired under altitude hypoxia (Chap. 1.1.1.4), while the standard $\dot{M}_{O_2}$ varies with what Brody (1945) called the "metabolically effective body weight." For the physiological significance of the parameters of the allometric relationship between standard $\dot{M}_{O_2}$ and body mass, see Heusner (1982a, b, 1983).

2.3.1.2 Increased Requirement for O_2

In newborn and small adult mammals it has been reported that exposure to even moderate hypoxia results in lowering of the oxygen uptake, heat production, and body temperature (Hill 1959; Rosenmann and Morrison 1974; Altland and Rattner 1981; Hayashi and Nagasaka 1981). Whether the metabolic demand was "standard," i.e., minimal, in these experiments is a crucial point. Hill (1959) stressed that "... the conflicting results reported for small mammals, and the apparent difference between them and larger mammals, can both be explained by

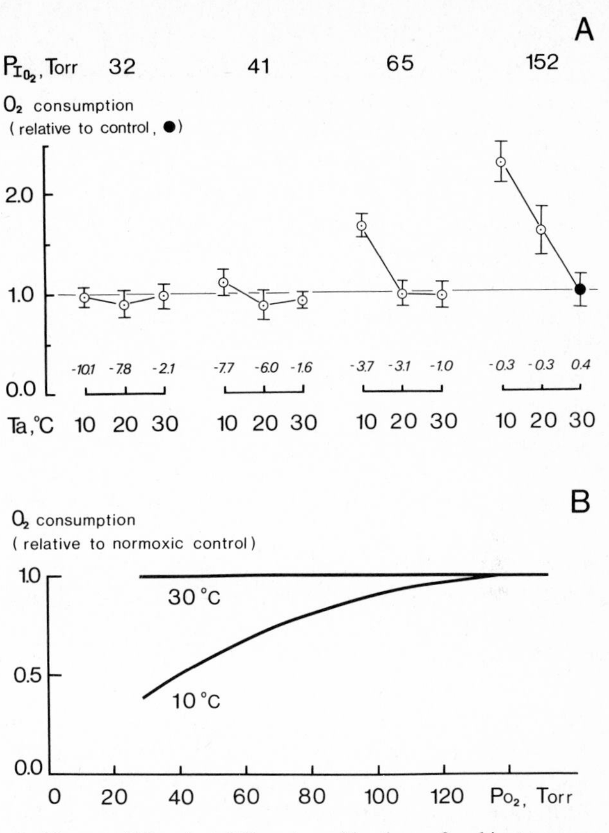

Fig. 2.4. Resting O_2 consumption of white rats (Wistar) at different combinations of ambient temperatures and partial pressures of O_2. (After Arieli et al. 1977). *A* Mean values ($\pm$ SD; n = 6) of the O_2 consumption were measured at rest in 12 experimental conditions of temperature (30°, 20°, and 10 °C) and P_{O_2} (152, 65, 41, and 32 Torr in ambient dry gas), and have been expressed as relative to the normoxic value at 30 °C (*black dot*). *Italics* changes in rectal temperature in the course of each experiment (°C); rats failed to maintain their normal body temperature when exposed to hypoxia, even at 30 °C ambient temperature. *B* Schematic representation derived from *A*, and showing that, when the level of oxygenation decreased acutely (30–60 min), the resting O_2 consumption did not deviate from normoxic control when measurements were made at thermoneutral temperature (30 °C), but was greatly depressed under exposure to cold (10 °C), when the O_2 demand for thermogenesis increased (in *A* compare the values at 10° and 30 °C for P_{O_2} = 152 Torr)

the hypothesis that hypoxia leads to lowered oxygen consumption only if the metabolic rate was above the basal level."

Figure 2.4 shows that, in normoxia ($P_{I_{O_2}}$ = 152 Torr), exposure to ambient temperature below that of thermoneutrality (20°–10 °C) caused an increase of the oxygen uptake in the animals studied by Arieli et al. (1977). The increased O_2 consumption became less pronounced at P_{O_2} of 65 Torr, and was no longer observed at P_{O_2} of 41 and 32 Torr. When compared with normoxic controls, the cold-induced O_2 consumption was clearly depressed by hypoxia; at 10 °C, the critical O_2 partial pressure was elevated, and close to the normoxic range (Fig. 2.4B). In comparison, Arieli et al. (1977) found the critical P_{O_2} at 10 °C to be lower than 65 Torr in mole rats, fossorial rodents adapted to their burrows' atmosphere where hypoxia (and hypercapnia) are likely to prevail. This observation suggests some functional adaptations in mole rats as compared with laboratory rats.

Similarly, rodents native to high altitudes have been reported to maintain oxygen consumption at lower P_{O_2}'s than their sea-level relatives (Bullard and Kollias 1966). Thus, at oxygen demand corresponding to 3.8 times the standard level, Rosenmann and Morrison (1975) found the critical P_{O_2} to be 110 Torr for highland rodents as compared with 122 Torr for their lowland relatives. This suggests a more efficient O_2 conductance in the highland animals; in terms of altitude the difference between groups was equivalent to about 900 m. On the other hand, comparing two groups of quail, one kept at sea level and the other acclimated for 6 weeks at 6.1 km altitude in a hypobaric chamber, Weathers and Snyder (1974) noted that, during acute progressive hypoxia at 5 °C, the mean

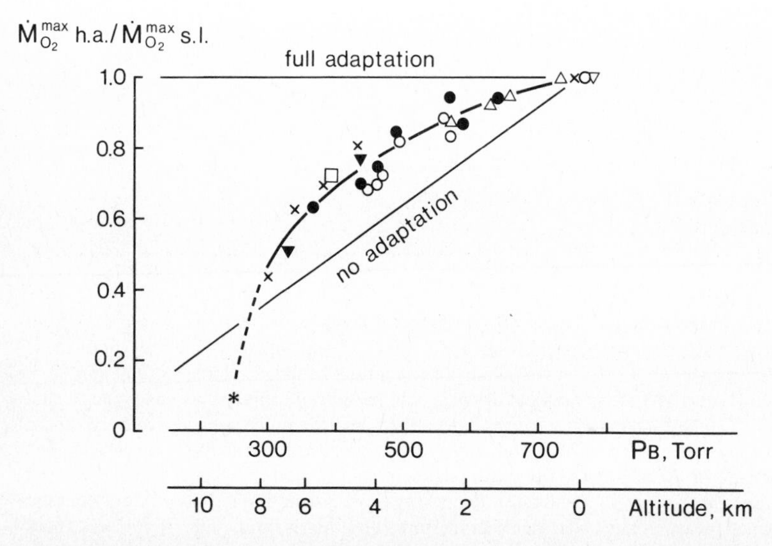

Fig. 2.5. Maximal oxygen consumption ($\dot{M}_{O_2}^{max}$) at high altitude (*h.a.*) relative to sea-level value (*s.l.*) as a function of barometric pressure (PB) or altitude. A free-hand curve has been drawn through the following experimental data, all obtained in man at various stages of exposure to high altitude: x, studies carried out on the 1960–1961 Himalayan scientific and mountaineering expeditions (Pugh 1964); ▼ and ▽, Peruvian natives studied respectively at *h.a.* and *s.l.* (Elsner et al. 1964); ○ and ●, data from various authors obtained respectively in acute hypoxic stages and after various periods of acclimatization (after Åstrand and Rodah 1970, p 573); □, lowlanders adapted to 5.35 km in the course of a 4-month expedition to Mt. Everest (Cerretelli 1976); △, near-sea-level residents in a hypobaric chamber for 1–2 h (Squires and Busbirk 1982); *asterisk* hypothetical point at the summit of Mt. Everest (West and Wagner 1980). Note the progressive reduction of $\dot{M}_{O_2}^{max}$ with increasing elevations as compared to sea level. There is neither altitudinal threshold, nor clear difference between subjects acutely exposed to hypoxia and those acclimatized to it. The oblique line ("no adaptation") visualizes how much $\dot{M}_{O_2}^{max}$ would have dropped if there were no increase in the overall, maximal O_2 conductance of the gas exchange system; assuming that P_{O_2} in mitochondria is negligible, hence $\Delta p_x = P_{I_{O_2}}$ in Eq. (2.1), the line shows the changes in the $P_{I_{O_2}}$ *h.a.*/$P_{I_{O_2}}$ *s.l.* ratio as a function of PB. The horizontal line ("full adaptation") indicates constancy of $\dot{M}_{O_2}^{max}$ at the *s.l.* value irrespective of elevation. See text for further comments

oxygen consumption of both groups of birds decreased rapidly below a P_{O_2} of 80 Torr.

For maximal oxygen consumption ($\dot{M}_{O_2}^{max}$), that may be attained in the course of sustained, exhausting muscular activity, the bulk of information comes from studies on human beings, confirmed by a few experiments on dogs. The data and solid curve in Fig. 2.5 show that $\dot{M}_{O_2}^{max}$, in contrast with the standard $\dot{M}_{O_2}$, but like the cold-induced O_2 demand, is strongly O_2-dependent: it declines with ascent to high elevations. A significant 5–7% change is already observed at 1.2–1.5 km (Squires and Buskirk 1982), and the rate of change increases rapidly above 5 km. Furthermore, it is practically impossible to find any difference in maximal performance between subjects acutely exposed to hypoxia and those acclimatized to it (Cerretelli 1980). Also, since both the energy cost for performing a given level of work and the proportional relationship between workload and oxygen consumption remain unchanged by altitude, the mechanical efficiency is the same for the highland as for the lowland natives (Buskirk 1978).

In Fig. 2.5, conditions of O_2-independence are shown by the horizontal "full adaptation" line. In such theoretical conditions, G_{O_2}, the overall O_2 conductance of the gas-exchange system, would have to increase at high altitude to such an extent that $\dot{M}_{O_2}^{max}$ would not depart from its sea level value despite the decrease of P_{O_2} in the ambient medium. Contrarily, the oblique "no adaptation" line in Fig. 2.5 visualizes how much $\dot{M}_{O_2}^{max}$ would decline when ascending to higher elevation if there were no compensatory changes in G_{O_2} from its sea-level value. Obviously, the experimental data are above the oblique line, thus indicating that G_{O_2} increases at high altitude. The magnitude of such an adaptive change in G_{O_2} can be appreciated by the vertical distance between the oblique "no adaptation" line and the experimental data shown in Fig. 2.5. This magnitude increases progressively from sea level up to about 5 km. It declines thereafter, indicating that there is deterioration rather than adaptation.

2.3.2 Anaerobic Metabolism

The rate of energy expenditure, $\dot{E}$ in Eq. (2.7), may exceed the oxygen consumption, $\dot{M}_{O_2}$, in those circumstances which, in vertebrates, correspond to unsteady states: (1) at the initiation of exercise, even when submaximal, until oxygen delivery systems adjust to oxygen requirements, i.e., during the transient constitution of an "oxygen debt," (2) during bursts of intense, supramaximal activity, (3) when the ambient P_{O_2} decreases down to a value below that of the critical P_{O_2} at which the overall O_2 conductance is no longer able to meet the oxygen demand. In these circumstances, organisms rely on anaerobic metabolism, at least in part; then, $\dot{M}_{AN}$ in Eq. (2.1) is no longer negligible.

Various biochemical reactions can be involved in anaerobic metabolism (Chap. 6). Those that yield products such as succinate and alanine appear to play a minor role. Classically, the most important reaction is fermentative glycolysis, in which the degradation of glycogen or glucose to lactate yields ATP without the need for oxygen; this is the lactic mechanism. Yet the splitting of ATP and phosphocreatine (PC) transiently yields energy with neither O_2 consumption nor lactate production; this is the alactic mechanism.

2.3.2.1 Alactic Mechanism

This mechanism contributes significantly to the energy requirement [$\dot{E}$ in Eq. (2.7)] only in the very early phase of muscular exercise. Present knowledge is restricted to humans, as recently reviewed by di Prampero (1981). At high altitude, a single study indicates that the alactic mechanism is not affected by acute or prolonged (3 weeks) hypoxia, at least up to 4.5 km (di Prampero et al. 1982).

2.3.2.2 Lactic Mechanism

Technological difficulties preclude the exact contribution of the lactic mechanism in the energy expenditure to be assessed. Present knowledge, reviewed by Bennett (1978) for lower vertebrates and by di Prampero (1981) for man and mammals, comes from assays of blood or skeletal muscle, or in whole-body homogenates of lizards (Bennett and Licht 1972), salamanders (Gatz and Piiper 1979) or small rodents (Ruben and Battalia 1979).

On the other hand, the lactic mechanism has some drawbacks: (1) it is a temporary process, and the lactate formed must be subsequently removed; this necessary removal of lactate requires oxygen (Chap. 6) and time, which corresponds to a period of exhaustion; (2) it may result in the depletion of glycogen stores; (3) its energy yield is low compared to that of aerobic pathways; (4) when formed at high concentration, lactate can have disruptive effects on intracellular and blood pH, with resulting impairment of enzymatic function; (5) conversely, the balance between production and removal of the lactate formed may be affected by the maximal H^+ concentration that can be tolerated by cells before glycolysis is inhibited, and the H^+ concentration is in turn determined by the buffer characteristics of the blood and tissues.

Acute hypoxia does not seem to affect the lactic mechanism in human beings, whereas in chronic hypoxia (at 5.35 km elevation) both the body-buffering capacity of the subjects and the maximal lactate concentration in blood after strenuous exercise are about half the values found at sea level (Cerretelli 1980). Such a limitation of the lactate production, i.e., of the amount of energy released by anaerobic glycolysis, in those subjects chronically exposed to high altitude, increases the energetic deficit, owing to the reduced $\dot{M}_{O_2}^{max}$. On the other hand, when exercising in hypoxia, lactate is first produced at work loads consistently lower than in normoxia, i.e., the "anaerobic threshold" is lower (Cerretelli 1980).

Ectotherms are known to be highly dependent upon anaerobic metabolism during activity (Hochachka and Somero 1973). It seems that high-altitude species have not evolved anaerobic adaptations to hypoxia, at least as noted in sceloporine lizards (Bennett and Ruben 1975).

Chapter 3

Ventilatory Adaptations

As seen in Fig. 2.2, the magnitude of the step in P_{O_2} from ambient air (I) to alveolar gas (A) decreases in mammals at high altitude because they do hyperventilate, replacing more of the alveolar gas with freshly inspired air, and consequently elevating the O_2 partial pressure at the medium/blood interface. Such hyperventilation is of adaptive value, being functionally equivalent to a descent to a lower altitude (Rahn and Otis 1949).

As extensively documented in man and other mammals (Hurtado 1964; Kellogg 1968; Lenfant 1973; Pace 1974; Lahiri and Gelfand 1981; Dempsey and Forster 1982), the ventilatory response to hypoxia is a time-dependent process, which is mediated by the hypoxic stimulation of arterial chemoreceptors, increases within the acute and short-term period of hypoxic exposure (minutes to days), and persists in the long-term exposure (months, years). Much remains to be learned about other vertebrates, although some aspects of the process have been described in birds (Black and Tenney 1980; Bouverot 1978; Colacino et al. 1977), reptiles (White 1978), and fish (Holeton 1980; Randall 1982).

3.1 Major Features

Both the magnitude and rate of development of the ventilatory response to hypoxia vary among species, and depend on the severity and duration of hypoxic exposure. Some major features can be summarized as follows.

3.1.1 Strategies

Conceivably, when the O_2 availability decreases in the environment, there are two strategies for maintaining an adequate O_2 transfer to the blood. One is to propel more ambient medium through the ventilatory apparatus, the other is to extract more O_2 from that medium; a combination of the two strategies is also possible.

3.1.1.1 O_2 Extraction and Specific Ventilation. Basic Equations

When applied to oxygen and to animals in steady state in which the total oxygen uptake is fully supplied by ventilation of gill or lung, Eqs. (2.4) and (2.5 b) may be written together as follows (Dejours et al. 1970; see also Dejours 1981):

$$\dot{M}_{O_2} = \dot{V} \cdot (Cin_{O_2} - Cout_{O_2}) = \dot{V} \cdot \beta_{O_2} \cdot (Pin_{O_2} - Pout_{O_2}), \quad (3.1)$$

where $\dot{M}_{O_2}$ is the amount of oxygen taken up in steady state from the ambient medium per unit time, $\dot{V}$ the ventilatory flow rate, Cin_{O_2}, $Cout_{O_2}$, Pin_{O_2}, and $Pout_{O_2}$ the oxygen cencentrations and partial pressures in the ingoing and outgoing medium (air or water), and β_{O_2} the O_2 capacitance coefficient of that medium.

Dejours et al. (1970) pointed out that a form of Eq. (3.1) convenient for comparative purposes is:

$$\dot{M}_{O_2} = \dot{V} \cdot E_{O_2} \cdot Cin_{O_2} = \dot{V} \cdot E_{O_2} \cdot \beta_{O_2} \cdot Pin_{O_2}, \tag{3.2}$$

where

$$E_{O_2} = \frac{Cin_{O_2} - Cout_{O_2}}{Cin_{O_2}} = \frac{Pin_{O_2} - Pout_{O_2}}{Pin_{O_2}} \tag{3.3}$$

is the ratio of the amount of O_2 used to the amount of O_2 available, i.e., the O_2 extraction coefficient, also designated "percent O_2 utilization" by some investigators.

It is not possible to make a direct comparison of the ventilatory flow rate, $\dot{V}$, among animals without taking into account the intensity of the O_2 consumption, $\dot{M}_{O_2}$, which varies with species, size and activity. A valid comparison can be made, using the ratio $\dot{V}/\dot{M}_{O_2}$, which is the *$\dot{M}_{O_2}$-specific ventilation,* or the convection requirement of air or water per unit of oxygen uptake. This ratio, sometimes called respiratory equivalent, designates the volume of medium which must be passed over the gas-exchange surface to effect delivery of a unit quantity of oxygen. It is easily derived by normalizing Eqs. (3.1) and (3.2) for $\dot{M}_{O_2}$

$$\frac{\dot{V}}{\dot{M}_{O_2}} = \frac{1}{Cin_{O_2} - Cout_{O_2}} = \frac{1}{E_{O_2} \cdot Cin_{O_2}} = \frac{1}{E_{O_2} \cdot \beta_{O_2} \cdot Pin_{O_2}}. \tag{3.4}$$

Equation (3.4) states that the product $(\dot{V}/\dot{M}_{O_2}) \cdot E_{O_2}$ must increase when the O_2 availability ($Cin_{O_2} = \beta_{O_2} \cdot Pin_{O_2}$) decreases in the inhaled medium. At constant value of E_{O_2}, Eq. (3.4) also states that the curve describing the specific ventilation as a function of the O_2 availability must take the form of a hyperbola. In most cases, however, the experimental data do not fit closely on a hyperbola, because E_{O_2} does not remain constant.

3.1.1.2 Comparison Among Animals

Figure 3.1 (upper part) provides graphic solutions of Eq. (3.4) for various values of E_{O_2}, and displays selected data from the literature for air breathers (open symbols) and water breathers (closed symbols). The lower part of the figure is a composite abscissa which allows appreciation of the environmental O_2 availability from either the O_2 concentration or the O_2 partial pressure of the inhaled medium. The following observations can be made:

a) Shaded areas are meaningless: E_{O_2} cannot approach 1 too closely, because all oxygen cannot be extracted from the ingoing media; on the other hand, it apparently does not decrease below 0.1.

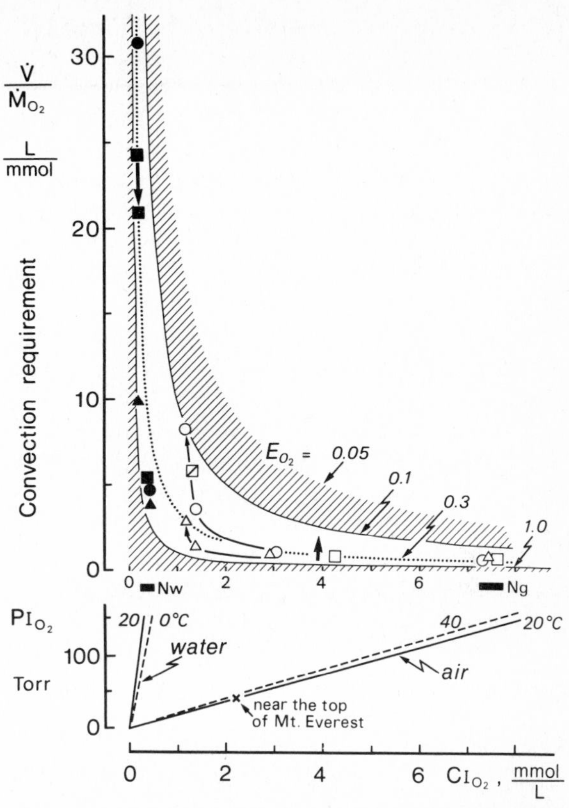

Fig. 3.1. Requirements for convection of water or air per unit of O_2 uptake (specific ventilation) as functions of the O_2 concentration and pressure of the external medium, water or air, in various water- and air-breathers. (After Dejours 1981). *Below*: linear relationships between the O_2 partial pressure (PI_{O_2}) and concentration (CI_{O_2}) in the inhaled water or air at various body temperatures; the reciprocal of the slope of each line corresponds to the O_2 capacitance coefficient of the medium (β_{O_2} in Table A 3). *Above*: specific ventilation, $\dot{V}/\dot{M}_{O_2}$, as a function of CI_{O_2}; *black dashes* on the *abscissa* indicate normoxic water (*Nw*) and normoxic air (*Ng*). *Curves* graphical solutions of the ventilatory O_2 convective equation [Eq. (3.4)] for various values of the O_2 extraction coefficient, E_{O_2}; shaded areas are meaningless (see text). *Open symbols* air-breathers; □, man native to low altitude exposed at 4.31 km elevation for 3 days (Grover 1963; *vertical arrow* drawn at 4 mmol · L^{-1} O_2 concentration visualizes the time-dependent process of ventilatory acclimatization to high altitude); ⧄, human natives (born above 3.7 km) exposed for short time to very high simulated altitude (Velasquez 1959); ○ and △, white Pekin duck and bar headed goose, respectively, acclimated for 1 month at the altitude of 5.64 km in a hypobaric chamber, then exposed acutely at various levels of normobaric hypoxia (Black and Tenney 1980). *Closed symbols* water-breathers; normoxia = 3 lower data points; normobaric hypoxia = 4 upper data points; ■, carp at 15 °C (Lomholt and Johansen 1979; *arrow* shows the changes in the ventilatory response observed after 1 month acclimation); ●, crayfish at 13 °C (Sinha and Dejours 1979); ▲, trout at 12 °C (van Dam 1938)

b) The unshaded area (between the curves drawn for E_{O_2} = 1 and 0.1) contains all the data points and can be viewed as the unavoidable path for the ventilatory response to changes in the ambient availability of O_2.

c) In normoxia (ambient P_{O_2} close to 150 Torr), the O_2 concentration is much higher in air than in water (compare Ng and Nw black lines on the abscissa).

This is the consequence of the very different O_2 capacitances of these media (Table A.3), and the cause which makes the specific ventilation, $\dot{V}/\dot{M}_{O_2}$, lower in air-breathers (open symbols, right) than in water-breathers (closed symbols, below), as stressed by Rahn (1966) and Dejours et al. (1970). Yet $\dot{V}/\dot{M}_{O_2}$ would be still greater in the aquatic animals if they had extracted O_2 from the inhaled medium as air-breathers do; in fact they extract two to three times more O_2 (Fig. 3.1 indicates that E_{O_2}, is around 0.7–0.8 in the aquatic animals, but 0.2–0.3 in the aerial ones). Such a high O_2 extraction by water-breathers is generally attributed to the efficient counter-current system of gills (Randall 1970, 1982; Shelton 1970).

d) In hypoxia, the data points for aquatic animals (closed symbols) range along the vertical arm of the unshaded area (in Fig. 3.1); $\dot{V}/\dot{M}_{O_2}$ increases dramatically and E_{O_2} falls to 0.3, or less. In contrast, the data points for air-breathers (open symbols) range along the horizontal arm of the unshaded area; from right to left they show two types of ventilatory responses: (1) $\dot{V}/\dot{M}_{O_2}$ and E_{O_2} both increase at first moderately with decreasing O_2 availability (the contracted ordinate of Fig. 3.1 makes the changes of $\dot{V}/\dot{M}_{O_2}$ hardly discernible), (2) only under severe hypoxia does $\dot{V}/\dot{M}_{O_2}$ increase abruptly and E_{O_2} decrease, much as in water-breathers; yet not all air-breathers behave in the same way (compare triangles with square and circle at about 1.25 mmol $\cdot$ L^{-1} O_2 concentration).

3.1.1.3 Interpretation

Equations (3.2) and (3.4) indicate that (1) at steady level of oxygenation, the specific ventilation, $\dot{V}/\dot{M}_{O_2}$, and the O_2 extraction coefficient, E_{O_2}, are inversely related, and (2) when going from normoxia to hypoxia, the product $(\dot{V}/\dot{M}_{O_2}) \cdot E_{O_2}$ must increase.

Considerable confusion, however, exists in the literature regarding E_{O_2} as it relates to hypoxia. This because the availability of ambient O_2 ($Cin_{O_2} = \beta_{O_2} \cdot Pin_{O_2}$) weights in both the numerator and denominator of the E_{O_2} ratio [Eq. (3.3)]. The numerator corresponds to the O_2 utilization, i.e., the amount of O_2 taken from each liter of ventilated medium; it is all-embracing, being affected by any change that alters the characteristics of the gas-exchange system, e.g., the match between ventilation and perfusion, the diffusive conductance of the respiratory surfaces, the physicochemical processes involved in the loading of blood with oxygen.

The following examples, based on Fig. 3.1, should remove difficulties; their functional implications will be examined in the next section.

a) E_{O_2} increases when going from normoxia to hypoxia if the characteristics of the gas exchange system are such, or change in such a way that the O_2 utilization [$Cin_{O_2} - Cout_{O_2}$; numerator of Eq. (3.3)] remains constant, or decreases proportionately less than the O_2 availability [denominator of Eq. (3.3)]. In the first case, there is no ventilatory response to hypoxia, according to Eq. (3.1); in the second case, a compensating, relatively small ventilatory response is necessary. The O_2 utilization remains constant in the remarkable bar headed geese studied by Black and Tenney (1980), which retain normal oxygen consumption, and deal so well with hypoxia that they do not need to hyperventilate until the altitude of

6.1 km (compare open triangles at 7.5 and 3 mmol · L^{-1} O_2 concentration in Fig. 3.1). Even at the simulated altitude of 10.7 km (open triangle at 1.4 mmol · L^{-1} O_2 concentration), these birds take an amount of oxygen from the inspired air that, although smaller than at sea level, is large enough to necessitate only a small increase in ventilation. At the two high altitudes, the O_2 extraction coefficient is greater than at sea level, but it is greatest at 6.1 km. Under more moderate hypoxia, most aerial animals exhibit ventilatory responses such that both $\dot{V}/\dot{M}_{O_2}$ and E_{O_2} increase at first; as hypoxic exposure is prolonged, however, $\dot{V}/\dot{M}_{O_2}$ increases further and E_{O_2}returns toward lower values (arrow at 4 mmol · L^{-1} O_2 concentration in Fig. 3.1; Sect. 3.1.4).

b) E_{O_2} remains constant when the O_2 utilization decreases in the same proportion as the O_2 availability; this requires the ventilatory response to hypoxia to be large enough (open circle at 3 mmol · L^{-1} O_2 concentration).

c) E_{O_2} decreases when the O_2 utilization declines proportionally more than the O_2 availability, in association with a larger increase in ventilation. This occurs in those air-breathers that are exposed to very severe hypoxia (upper open circles and squares in Fig. 3.1) and in almost all water-breathers (upper closed symbols).

3.1.1.4 Resulting O_2 Driving Pressure

One easily derives from Eq. (3.3) that

$$Pout_{O_2} = (1 - E_{O_2}) \cdot Pin_{O_2}. \tag{3.5}$$

At given partial pressure of inspired O_2 (Pin_{O_2}), hence elevation, Eq. (3.5) states that the lower the O_2 extraction coefficient (E_{O_2}; hence the greater the specific ventilation), the higher the O_2 pressure head at the gas-exchange surface, at least in so far as $Pout_{O_2}$ can be considered as a valid index of that O_2 pressure head.

In air-breathers, such a valid index can easily be taken from the P_{O_2} value in the gas vented from the lungs in late expiration, after the dead space has been washed out (end-tidal, or alveolar gas; Chap. 2.2.1 a). Most of the data points shown in Fig. 3.1 (open symbols) suggest that the O_2-driving pressure at the pulmonary gas exchange surface does increase under hypoxia, due to hyperventilation and decrease in E_{O_2}, either acutely (far left open circles and squares) or chronically (arrow at 4 mmol · L^{-1} O_2 concentration). In contrast, the bar headed geese at the altitude of 6.1 km (open triangle at 3 mmol · L^{-1} O_2 concentration) do not hyperventilate, but increase E_{O_2}, and therefore appear to tolerate rather low O_2 driving pressure. Since they retain normal oxygen uptake (Black and Tenney 1980), such a low driving P_{O_2} has to be compensated elsewhere along the O_2 transport path. The evidence at hand indicates that fossorial mammals may behave in the same way, since they do not hyperventilate until the development of very severe hypoxia (Arieli and Ar 1979). On the other hand, the llama, a high-altitude native camelid, exhibits a ventilatory response to hypoxia which is much like that of other mammals (Brooks and Tenney 1968). Clues to these different strategies will be studied later.

Regarding water-breathers and their "open" ventilatory system (Chap. 2.2.1 d), a valid index of the O_2 pressure head at the gas exchange surface is not easily available. Nevertheless, the last column of Fig. 2.3 (bottom) allows an intuitive understanding that a ventilatory response to hypoxia large enough to lower E_{O_2} will, according to Eq. (3.5), raise P_{O_2} in the expired water ($Pout_{O_2}$) and shift the O_2 partial pressure profile upward along the gill's exchange surface, hence raising the mean driving O_2 pressure at the water/blood interface. Such an adaptive response occurs in water-breathers (Shelton 1970; Itazawa and Takeda 1978; upper filled symbols in Fig. 3.1). Whether secondary improvement of the efficiency of the internal O_2 transfer (increased E_{O_2}) may allow the hypoxia-acclimated water-breathers to maintain O_2 uptake at a smaller ventilation than in the early phase of hypoxic exposure is not well documented. Such a delayed, compensatory mechanism, however, apparently occurred in the carp studied by Lomholt and Johansen (1979; arrow between filled squares in Fig. 3.1) before and after acclimation to hypoxic water for 1 month (P_{O_2} = 30 Torr).

3.1.2 Tactics

Full understanding of the process of ventilatory adaptation to hypoxia requires knowledge of several fundamental variables in addition to those already considered. As yet, this is not the case in many studies, a defect that renders their interpretation difficult or uncertain.

3.1.2.1 Effective Ventilation

From the total ventilatory flow rate ($\dot{V}$) considered so far, only a part, the effective ventilation ($\dot{V}eff$), is involved in the exchanges of O_2 and CO_2 molecules at the medium/blood interface. The other part, the dead-space ventilation ($\dot{V}_D$), does not exchange O_2 and CC_2, because it does not come into intimate diffusional contact with blood. The dead-space ventilation is equal to the product $f_R \cdot V_D$, where f_R is the ventilatory frequency, and V_D the volume of the physiological dead space. In air breathers, V_D can be viewed as the volume of the conducting airways plus the pulmonary spaces that are not perfused; in water breathers, it is the part of the stroke volume of water that comes in contact with unperfused gill lamellae and/or is shunted away from perfused lamellae.

The following equation can be written

$$\dot{V}eff = \dot{V} - f_R \cdot V_D , \qquad (3.6)$$

which states that, at given ventilatory output ($\dot{V}$), the lower the product $f_R \cdot V_D$ (or dead-space ventilation, $\dot{V}_D$), the greater the effective ventilation ($\dot{V}eff$). Conceivably, therefore, any decrease in f_R, or in V_D, or in both factors, will optimize the O_2 transfer from ambient medium to blood, and reduce the ventilatory needs. A decrease in the respiratory frequency is a matter of breathing pattern, which has been reported to contribute in the process of adaptation to high altitude (Bureau and Bouverot 1975). A decrease in the dead-space volume, as resulting from more even distribution of the flows of blood and ambient medium at the gas-exchange surface, may also contribute. For instance, recruitment of unperfused lamellae and/or more uniform distribution of the inspired stroke volume were possibly involved in the carp studied in prolonged hypoxia by Lomholt and Johansen (1979; arrow between solid squares in Fig. 3.1); if so the *effective* specific ventilation, $\dot{V}eff/\dot{M}_{O_2}$, may have been unaffected when the *total* specific ventilation, $\dot{V}/\dot{M}_{O_2}$, decreased. A combination of a direct myogenic response of gill vessels and release of adrenaline are possible mechanisms (Pettersson and Johansen 1982). In man and other mammals, better blood flow to some parts of the lungs, such as the apices, due to pulmonary hypertension, is also thought to improve oxygenation (Fishman 1976, Chap. 4.3.2.6).

Equation (3.6) also indicates that an important increase in the respiratory frequency may cancel the beneficial effect of increasing the total ventilation. This applies to the tachypneic response reported to occur in some animals deprived of arterial chemoreceptors. Such a response has no adaptive value, as will be seen below.

3.1.2.2 Breathing Pattern

In air- and water-breathers, the periodic motor act of pumping air or water is best described by the use of both its amplitude (the tidal or stroke volume, V_T) and period (T), or its inverse the ventilatory frequency (f_R $min^{-1} = 60/T$, with T in sec). Thus, the total ventilation, or minute volume is

$$\dot{V} = V_T \frac{60}{T} = V_T \cdot f_R \,. \qquad (3.7)$$

The corresponding plot of V_T against T and some exemplifying data points are shown in Fig. 3.2. It is important to differentiate

1. a steady ventilation ($\dot{V}$), as illustrated by any given oblique line, which may result from an infinity of combination between V_T and T,

2. the breathing pattern, which corresponds to a given point on a given isoventilation oblique line, and results from the control of breathing [2],

3. hyperventilation, which implies that this point jumps from one isoventilation line to a steeper one, and results from the control of ventilation [2].

Even in the so-called steady state, tidal volume, ventilatory period, as well as O_2 and CO_2 transfers across the medium/blood interface, are continuously varying around mean values (Dejours et al. 1966; Imbert et al. 1976). These breath-by-breath ventilatory oscillations appear to be affected differently in the various phases of high-altitude exposure, with more dramatic effects during sleep (Sect. 3.3.3). They may be less consistently noted in chronic than in acute hypoxic exposure (Brusil et al. 1980).

2 *The control of breathing* yields a given combination of amplitude (V_T) and period (T), which results from the activity of a functional unit, the ventilatory neuromechanical system, organized in a feedback loop: the so-called respiratory centers and the ventilatory pump, interconnected by motor pathways and by sensory afferent pathways from specific receptors in the ventilatory apparatus. *The control of ventilation* yields a change in the motor output of that functional unit, induced by humoral and/or neuronal stimuli which inform the respiratory centers from the ventilatory needs of the body (see Dejours 1981, for further details)

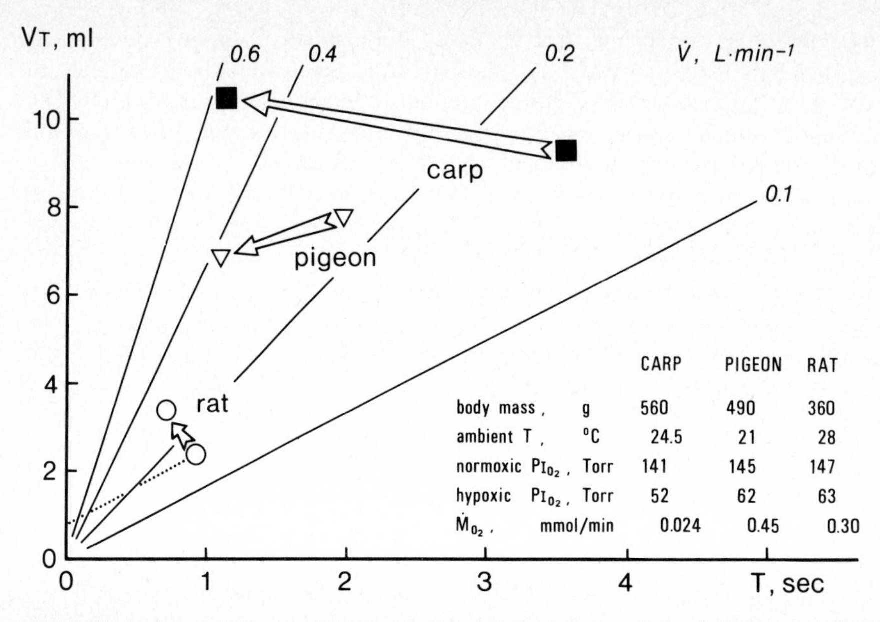

		CARP	PIGEON	RAT
body mass,	g	560	490	360
ambient T,	°C	24.5	21	28
normoxic P_{IO_2},	Torr	141	145	147
hypoxic P_{IO_2},	Torr	52	62	63
$\dot{M}_{O_2}$,	mmol/min	0.024	0.45	0.30

Fig. 3.2. Mean changes in the breathing pattern from normoxia to acute hypoxia in unanesthetized rat and pigeon (personal unpublished data) and carp (Itazawa and Takeda 1978). *Ordinate* tidal volume (V_T); *abscissa* ventilatory period (T); *oblique lines* isoventilation lines ($\dot{V}$). *Dotted line drawn through circle* (normoxic rat) and extrapolated to the left, intercepts the ordinate at 0.8 ml, a value close to the dead-space volume for rat; this line represents the various V_T-T combinations yielding constant effective ventilation ($\dot{V}eff$) (see Sect. 3.1.2.2)

Hyperventilation is commonly observed in transition from normoxia (arrow-tails in Fig. 3.2) to hypoxia (arrow-heads), when changes in both the amplitude and period appear to be involved. There are interspecies differences, however, and much remains to be learned about underlying mechanisms. This also holds for the neural mechanisms in the brain stem, that are responsible for the partitioning of the minute ventilation ($V_T \cdot f_R$) into a "drive" component (V_T/T_I) and a "timing" component (T_I/T), where T_I is the inspiratory time (Milic-Emili and Grunstein 1976; Gautier et al. 1982). Changes in the patency of the airway, mainly laryngeal (Bartlett 1979), as well as in the mechanics of the ventilatory pump (Gautier et al. 1982) may also provide an important influence in determining the breathing pattern of aerial vertebrates during hypoxia.

A tachypneic response to hypoxia may have no adaptive value; for example, that exhibited by those mammals and birds experimentally deprived of arterial chemoreceptors (Davenport et al. 1947; Miller and Tenney 1975; Bouverot and Sébert 1979). This is schematically illustrated for rats by the dotted line in Fig. 3.2, which intercepts the Y-axis at 0.8 ml, a value close to V_D, the expected dead-space volume (Stahl 1967); therefore, the general equation of the line is $V_T = V_D + aT$, where a is the slope. Dividing each member by T, and rearranging, leads to $a = \dot{V} - \dot{V}_D$, hence to $a = \dot{V}eff$ [see Eq. (3.6)]. This means that the dotted line represents the various V_T-T combinations yielding steady effective ventilation. As a data point, however, progresses down that dotted line, both V_T

and T decrease (fR increases) in such a way that the total ventilation increase (solid lines). Despite this tachypneic hyperventilation in chemodenervated animals, alveolar P_{O_2} will not return toward higher values, since the effective ventilation remains steady; similarly, alveolar P_{CO_2} will be unaffected, whereas it does decrease in intact animals, as considered next.

3.1.3 Associated Hypocapnia and Alkalosis

An effective increase in the specific ventilation means an equally increased ventilatory conductance for CO_2. The consequence is that the CO_2 partial pressure (and concentration) diminishes in the arterial blood and expired medium. This, in association with the roughly identical capacitances of air and water for CO_2

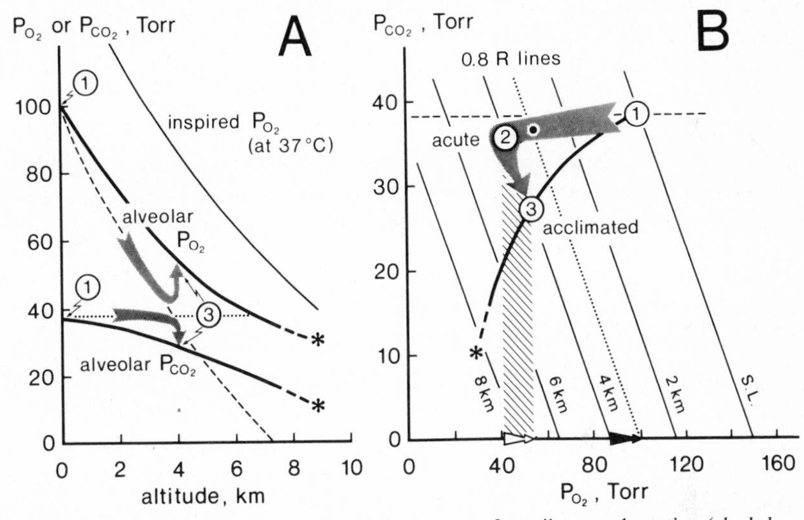

Fig. 3.3. Alveolar gas composition (*thick curves*) and the process of ventilatory adaptation (*shaded arrows*) at various altitudes above sea level. *A*: O_2 or CO_2 partial pressure in the inspired air (saturated with water vapor at 37 °C) and alveolar gas as function of altitude. Note the drop of inspired P_{O_2} with increasing elevations (*thin continuous line*). The alveolar P_{O_2} would drop in the same proposition if there was no ventilatory response to hypoxia (*dashed line*), hence no change in the alveolar P_{CO_2} (*dotted line*). Note the hyperventilatory decrease of P_{CO_2} and gain in P_{O_2} in the alveolar gas of humans sojourning at the various altitudes (*thick lines* redrawn from the data compiled by Pace, 1974); *stippled arrows* and *circled numbers* refer to the time-dependent process of ventilatory adaptation to hypoxia that is explained in *B*. *B*: P_{CO_2} vs. P_{O_2} diagram. (After Rahn and Otis, 1949). Each R-line originating at a P_{CO_2} value of 0 and a given P_{O_2} value corresponds to all the possible combinations of O_2 and CO_2 in the ventilated air at the altitude indicated, for a respiratory quotient of 0.8. *1* normoxia; *2* acute hypoxia (e.g., 1 h); *3* chronic hypoxia (e.g., days or weeks). *Thick curve* man acclimatized to various altitudes (this curve results from the combination of the two thick curves shown in *A*). *Stippled arrow* path that the alveolar gas composition takes in the process of ventilatory adaptation to the altitude of 4 km (this arrow is simply the combination of the two stippled arrows shown in *A*); the process of ventilatory "acclimatization" to that altitude (2→3) yields a gain in alveolar P_{O_2} which is visualized by the *hatched area* and *white arrow* on the abscissa; this gain is equivalent to a descent from 4 to 3 km (*black arrow* on the abscissa); for purposes of comparison, the alveolar gas composition during *acute* exposure to that *lower altitude* is shown by the *black dot* on the dotted R line = note the drop of P_{CO_2} when point 3 is compared with this black dot. *Asterisks* (*A* and *B*), hypothetical point at the summit of Mt. Everest (West and Wagner 1980)

(Table A.3), is the reason for the large difference in arterial P_{CO_2} between mammals and fish (40 versus 1–4 Torr; Rahn 1966). This is also the reason for the hypocapnia observed in the course of hypoxic exposure.

In Fig. 3.3 A, the dotted and dashed lines have been drawn assuming lack of ventilatory response to hypoxia and constant metabolic rate; they indicate that P_{CO_2} would remain steady in the alveolar gas, whereas P_{O_2} would drop in the same proportion as the inspired P_{O_2} drops as a function of altitude. The experimental data (thick, continuous lines) clearly depart from these hypothetical dashed and dotted lines, indicating that the higher the altitude, the greater the ventilatory response to hypoxia, the more closely the O_2 partial pressure in the alveolar gas approaches that in the inspired medium, and the lower is alveolar P_{CO_2}. The stippled arrows visualize this adaptive process, which is best shown using an O_2-CO_2 diagram (Rahn and Fenn 1955). As an example in Fig. 3.3 B, the white arrow on the abscissa underlines the gain of P_{O_2} that results from the ventilatory adaptation to the altitude of 4 km. The tribute paid is a proportional decrease in P_{CO_2}. Likewise in the arterial blood, P_{O_2} becomes higher, and P_{CO_2} becomes lower than it would have been without hyperventilation (Fig. 3.4 B). The decrease of P_{CO_2} in blood is accompanied at first by an elevation of the pH value (Fig. 3.4 C). As hyperventilation occurs, blood $[HCO_3^-]$ falls, primarily due to buffering by hemoglobin; subsequently, bicarbonates are excreted from the blood by the kidney (Haldane et al. 1919), and the pH returns, at least partially, toward control. Similar changes in the pH of cerebrospinal fluid have been reported (review by Dempsey and Forster 1982). Present knowledge results essentially from the numerous studies on mammals conducted since the pioneer report by Dill et al. (1937). Scarce, but similar data have been reported in birds (Besch et al. 1971; Bouverot et al. 1976; Lutz and Schmidt-Nielsen 1977) and fish (Itazawa and Takeda 1978).

3.1.4 Time Course

A growing body of evidence in aerial vertebrates indicates that, if the exposure to high altitude or hypoxia is more than momentary, the magnitude of the ventilatory response progressively increases, as do the associated changes in P_{O_2} and P_{CO_2} at the respiratory surface and in arterial blood. The process takes place over a period of hours or days, depending on the animal species, and stabilizes thereafter. Yet, other factors may render difficult the interpretation of some experimental findings. To put the problem in perspective, a recent experiment on the laboratory rat (an animal in which many aspects of metabolism have been detailed; Chap. 6) will be examined first.

3.1.4.1 A Study on Rats

Figure 3.4 A shows data obtained in conscious rats exposed for 2 weeks at the simulated altitude of 4.3 km in a hypobaric chamber (inspired P_{O_2} = 85 Torr). It may be seen that the specific total ventilation, $\dot{V}/\dot{M}_{O_2}$, increased rapidly during the first hour of hypoxia, then rose more slowly before stabilizing after about 1 day. In

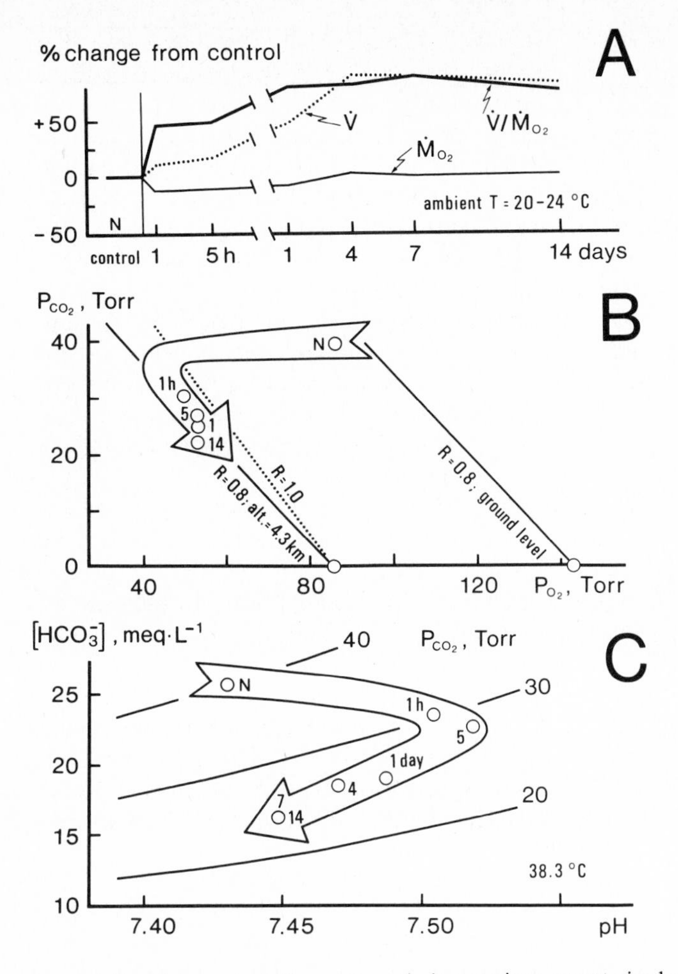

Fig. 3.4. Ventilatory adaptation to hypoxia in conscious, unrestrained rats. (Data from Olson and Dempsey 1978). All measurements were made in a plethysmograph under normobaric conditions (PB = 735 Torr). A total of 69 albino rats were studied first in normoxia (PI_{O_2} = 143 Torr), then at one or more time points of hypoxic exposure (PI_{O_2} = 85 Torr). For hypoxic exposure of 1 and 5 h, the rats were kept under normobaric conditions in an atmosphere of 12–12.5% O_2; for more prolonged hypoxic exposure, they were placed in a hypobaric chamber operated at PB = 450 Torr (4.3 km altitude). *A*: Time course of the relative changes in oxygen uptake ($\dot{M}_{O_2}$), minute ventilation ($\dot{V}$), and specific ventilation ($\dot{V}/\dot{M}_{O_2}$) for up to 2 weeks of hypoxia. These changes are expressed as percent of normoxic control (*N*). Note that the ambient temperature was 20°–24 °C, i.e., below the thermoneutral range for rats; it may be inferred that the measured $\dot{M}_{O_2}$ was not standard (Chap. 2.3.1.2). *B*: Arterial blood P_{O_2} and P_{CO_2} in normoxia (*N*) and at the various time points of hypoxic exposure indicated by *numbers* at each point. Although no alveolar value is plotted, the oblique R lines drawn may help in understanding the adaptive process that is visualized by the *arrow*. *C*: Acid-base status of arterial blood on a [HCO_3^-] vs. pH diagram, in normoxia (*N*) and at the periods of hypoxic exposure indicated by *numbers* at each point. *Arrow* visualizes the adaptation pathway: first, uncompensated respiratory alkalosis from *N* to 1 and 5 h; second, the arterial pH returns toward normal, due to partial metabolic compensation

contrast, the minute volume, $\dot{V}$, continued to increase for the initial 4 days. The difference between these two time courses relates to the transient reduction of the oxygen consumption, $\dot{M}_{O_2}$, in the early phase of hypoxia.

Why, in this study, did $\dot{M}_{O_2}$ fall transiently, whereas in most other mammals the standard oxygen consumption is reported to remain unaffected by equivalent hypoxia (Chap. 2.3.1)? Was the energy expenditure reduced, due to a decrease in the food intake (Westerterp 1977), which possibly was associated with the early phase of hypoxic exposure (Ettinger and Staddon 1982)? Was the metabolic demand above the standard level, and the overall O_2 conductance of the gas exchange system transiently unfitted to the O_2 needs? The latter possibility is supported by the fact that, in the experiment under consideration, the ambient temperature was 20°–24 °C, hence below the thermoneutral range for the rat (30°–34 °C); consequently, $\dot{M}_{O_2}$ rose in normoxia 40–50% above the standard level, due to the increased thermogenesis which has proved to be sensitive to hypoxia (Chap. 2.3.1.2).

Since only the changes in $\dot{V}/\dot{M}_{O_2}$ are easily interpreted (Sect. 3.1.1), Fig. 3.4 A suggests that the process of ventilatory adaptation was completed in the rats within the first 24 h of exposure to high altitude.

The process of ventilatory adaptation to hypoxia can also be documented by considering the changes in arterial P_{O_2} and P_{CO_2} (Fig. 3.4 B). The following observations can be made:

1. The decrease in inspired P_{O_2}, which occurred when going from near sea level to the simulated altitude of 4.3 km, can be appreciated by comparing the two data points on the abscissa;
2. If there were no ventilatory response to hypoxia, P_{CO_2} in the arterial blood would have remained at 40 Torr (on the horizontal line through N) and arterial P_{O_2} would have dropped approximately to 35 Torr, on the left of the oblique solid line (R = 0.8) originating at 85 Torr on the abscissa (Rahn and Fenn 1955).
3. In fact, due to the hypoxic ventilatory response, Pa_{O_2} decreased only to 50 Torr at 1 h of hypoxia, increased to 52 Torr after 5 h, and persisted at this level throughout of the remaining 14 days of hypoxic exposure; the gain in arterial P_{O_2}, therefore, was 52–35 = 17 Torr;
4. The tribute paid was a 12 Torr decrease of Pa_{CO_2} (Fig. 3.4 B) and a shift to highly alkaline pH at 5 h of hypoxia (Fig. 3.4 C). The progressive return of arterial pH toward control value between 5 h and 14 days of hypoxia (Fig. 3.4 C) does not proceed from the ventilatory adaptation itself, but from the partial compensation of the initial respiratory alkalosis by the secondary metabolic acidosis related to bicarbonate excretion from the blood by the kidneys.

3.1.4.2 The P_{O_2} Criterion

That the hyperventilation observed during hypoxia is adaptive in making the O_2 driving pressure at the gas-exchange surface increase (Rahn and Otis 1949), can be seen in Fig. 3.4 B, but is best visualized in Fig. 3.3 B. Thus, for instance, in a human exposed to the altitude of 4 km, the alveolar P_{O_2} would be 39 Torr in the absence of hyperventilation (intercept of the horizontal, dashed line drawn through the normoxic point, 1, and the oblique line originating at 86 Torr P_{O_2}).

In fact, due to the hyperventilatory response to hypoxia, which develops progressively, the alveolar P_{O_2} does rise to 42 Torr within the first hour (point 2), and to 54 Torr after a few days (point 3), thereafter stabilizing.

The secondary increase, from point 2 to point 3, is sometimes referred to as "ventilatory acclimatization." Its beneficial effect, in the example under consideration, is a gain of $54 - 42 = 12$ Torr in alveolar P_{O_2} (white arrow on the abscissa). In fact, the overall hyperventilatory process, yields a $54 - 39 = 15$ Torr increase of P_{O_2} at the gas exchange surface, a considerable advantage indeed: the subject has, as it were, descended from 4 km down to 3 km (black arrow on the abscissa). The tribute paid is hypocapnia (alveolar P_{CO_2} dropped 12 Torr) and related alkalosis.

3.1.4.3 Interspecies Differences

The magnitude of the ventilatory response to hypoxia in the early and secondary phases of hypoxic exposure differ among the few species so far studied (see Dempsey and Forster 1982). For purpose of comparison, a ΔP_{CO_2} vs./ΔP_{O_2} representation can be derived from the P_{CO_2}-P_{O_2} diagram of Rahn and Fenn (1955).

Figure 3.5 is such a representation of ΔP_{CO_2}-ΔP_{O_2}, comparing the P_{O_2} and P_{CO_2} changes in the alveolar gas of men to the P_{O_2} and P_{CO_2} changes in arterial blood of awake dogs and rats, when these mammals were exposed to prolonged similar hypoxia at an ambient temperature within the thermoneutral range. In each species, point 1 refers to control normoxic conditions, while points 2 and 3 show respectively the initial and final changes observed in the course of the hypoxic exposure. The dashed lines in Fig. 3.5, redrawn from Rahn and Otis (1949), visualize the changes described by these authors in man, unacclimatized (curve 1) and acclimatized (curve II) at various altitudes.

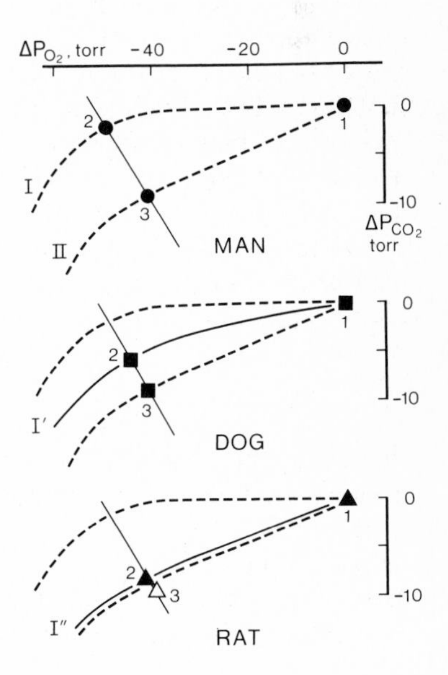

Fig. 3.5. Changes in the tensions of alveolar gas of man or arterial gas of dog and rat exposed to hypoxia either acutely point *2*, lines I, I', and I'') or chronically (point *3*, line II). The ΔP_{O_2}–ΔP_{CO_2} coordinates show the differences from the normoxic control values at the origine (point *1*). Conscious man and dog breathed ambient air at about 4000 m in an altitude chamber; the awake rat breathed a low O_2 gas mixture, resulting in the same inspired P_{O_2} (90 Torr) as in man and dog. *Dashed lines* I and II from Rahn and Otis (1949). *Solid lines* I' and I'' refer to unacclimatized awake dog (Bouverot et al. 1973) and rat (unpublished) respectively. Pathways *1-2-3* visualize the process of ventilatory adaptation in each species. Point *3* for the rat is an open symbol because a real altitude experiment was not made; the ambient temperature was 18°–20 °C for dogs, and 29°–30 °C for rats

a) Acute hypoxia: The upper part of Fig. 3.5 shows that in man P_{CO_2} does not fall until P_{O_2} decreases about 40–50 Torr (compare points 2 and 1), hence indicating that ventilation of unacclimatized man does not immediately increase when he is acutely exposed to moderately high altitude. By contrast, the middle and lower parts of Fig. 3.5 reveal that the unacclimatized dog and rat hyperventilate when acutely exposed to a similar hypoxia: at points 2, measurements made within 30 min after the beginning of hypoxic exposure, P_{CO_2} fell about 5 Torr in dogs, and 8 Torr in rats. This indicates a rapid onset of hypoxic hyperventilation in these animals, as opposed to man, and a greater acute ventilatory response in rat than in dog. These interspecies differences possibly relate to differences in body mass. They apparently relate to the inequal magnitudes of the hypoxic chemoreflex mechanism (Sect. 3.2.1.3), which has been found to contribute for about 15%, 35%, and 50% of the normoxic minute volume in man, dog and rat, respectively (Bouverot 1976).

b) Prolonged hypoxia: Points 3 of the upper and middle parts of Fig. 3.5 show that the overall changes in P_{O_2} and P_{CO_2} are similar in man and dog exposed to the same altitude, but the time required for completion of the process of ventilatory adaptation is quite different, since point 3 was reached after 2–3 days of hypoxia in man, as compared with 3 h in dog. On the other hand, in the rat exposed for several hours to an equivalent hypoxia (lower part of Fig. 3.5; point 3), the ultimate changes in P_{O_2} and P_{CO_2}, which were quantitatively similar to those in man and dog, were observed within 45–60 min, and in fact were not so different from those observed upon acute exposure to hypoxia.

c) Lifelong hypoxia may result in an attenuated (blunted) ventilatory response to hypoxia in human beings, born and raised at high altitude (Lefrançois et al. 1968; Severinghaus 1972). The underlying mechanisms are not understood yet, but the functional consequence may be viewed in Fig. 3.3 B or Fig. 3.5 (top) as making point 3 move back part way toward point 2; it may even fully return to point 2 in some patients suffering from chronic mountain sickness and polycythemia (Severinghaus et al. 1966a; Lahiri 1968). No satisfactory animal model was available until Tenney and Ou (1977b) showed that cats raised at high altitude may also exhibit a "blunted" ventilatory response to hypoxia. Such a "blunting" phenomenon is not shared by llamas (Brooks and Teney 1968) or yaks (Lahiri 1972) which, however, are native high altitude animals. On the other hand, steers (Grover et al. 1963) and ponies (Orr et al. 1975) develop at first a normal ventilatory adaptation to high altitude, with accompanying hypocapnia, but, by the end of a week only, a relative hypoventilation occurs, with a rise of P_{CO_2} in the arterial blood. The significance and mechanisms of these interspecific differences are unclear.

3.2 Mediation

Current knowledge results from studies essentially conducted in man and in a limited number of mammals and birds. Information for other vertebrates is virtually lacking. It is generally agreed that, in the acute phase of hypoxia, the early hyperventilation results from the hypoxic stimulation of arterial chemoreceptors. In contrast, the mediation of the further, time-dependent changes in ventilation remains controversial and unresolved. The evolution of ideas is provided in the reviews written by Kellogg (1968, 1977); further references are available in Dempsey and Forster (1982).

3.2.1 Hypoxic Chemoreflex Drive

The concept that the arterial chemoreceptors (often designated peripheral chemoreceptors) are responsible for the hyperventilation observed in acute hypoxia originates from the work of Heymans and associates in the late 1920's (references in Heymans and Neil 1958). Half a century later, we still do not know the mechanism by which the chemoreceptors are excited by hypoxia, although a considerable amount of work has been devoted to this problem.

Information about the structure, mechanisms of excitation, and functions of arterial chemoreceptors can be found in many proceedings and review articles (e.g., Torrance 1966; Biscoe 1971; Paintal 1976; Acker et al. 1977; Biscoe and Willshaw 1981; Lahiri and Gelfand 1981). Yet most of this voluminous information results from studies on cat's carotid body. What happens in the "chemoreceptor organ black box" will be ignored in the following analysis.

3.2.1.1 Definition

The hypoxic chemoreflex drive is that mechanism by which the partial pressure of oxygen in the arterial blood stimulates the arterial chemoreceptors, and, through their action, increases the activity of the respiratory centers, and consequently that of the ventilatory pump.

a) The reflex arc is formed by the chemoreceptors, the afferent fibers to the nervous centers, these centers, and the motor pathways to the ventilatory pump. In mammals there are two bilateral sets of arterial chemoreceptors (i.e., four sets of afferent fibers): the carotid bodies innervated by the sinus branches of the glossopharyngeal nerves, and the aortic bodies innervated by the aortic nerves which are often included in the sheet of the vagi. The excitatory effect of arterial chemoreceptors, known to be at work in normoxia and to increase in hypoxia, is processed centrally and transmitted within seconds to the muscles of ventilation.

Discharging arterial chemoreceptors afferents not only stimulate the phasic activity of respiratory neurons, but they may cause a tonic "inspiratory shift" in the pattern of respiratory motoneuron activation (Sears et al. 1982). They may thereby contribute to the hypoxic increase in lung volume reported in mammals (Bouverot and Fitzgerald 1969), which may be advantageous for pulmonary gas exchange by increasing the alveolar gas reservoir and expansion of respiratory surface area. The arterial chemoreceptors apparently affect also the activity of the motoneurons controlling the patency of pharyngeal airway and, therefore, the total load on breathing (Bartlett 1979; Brouillette and Thach 1980).

b) The "oxygen" or "hypoxic" stimulus is classically defined as a factor of control of breathing, related to the partial pressure of oxygen in the arterial blood, Pa_{O_2} (Dejours 1962). Since the chemoreceptor activity increases when Pa_{O_2} decreases, and provokes reflexly an increase of ventilation, the "O_2 stimulus" is better called a "hypoxic stimulus;" both terms, however, will be used exchangeably.

There is overwhelming evidence that arterial P_{O_2} is the appropriate index of stimulus, and determines the response of the chemoreceptors (literature quoted above). However, at the cellular level, O_2 chemoreception proceeds from oxidative phosphorylation (Mulligan et al. 1981), and it is most certainly the receptor P_{O_2} that is sensed. Yet the relation of tissue P_{O_2} at the receptor to the activity of that receptor remains to be worked out, as do the relationships between local blood flow, tissue P_{O_2}, and total blood flow of the chemoreceptors organs (see Acker et al. 1977). On the other hand, some results indicate that aortic chemoreceptors sense O_2 concentration as well as P_{O_2} in the arterial blood, as also the carotid body when it has been sympathectomized (Hatcher et al. 1978; Lahiri 1980).

In connection with the view that intact oxidative phosphorylation is necessary for the expression of O_2 chemoreception, it is worth noting that NaCN (sodium cyanide which inhibits the uptake of oxygen by cytochrome oxidase) is widely used in experiments to stimulate arterial chemoreceptors.

c) Threshold: Accepting Pa_{O_2} as the appropriate index of stimulus, the threshold for stimulation of arterial chemoreceptors is above the normoxic value of P_{O_2} in the arterial blood: the activity in single afferents from the carotid body of the cat increases hyperbolically when Pa_{O_2} is lowered below 200 Torr approximately (see Biscoe and Willshaw 1981).

d) Interactions between arterial chemoreceptors and the other factors of control of ventilation are important. It is generally agreed that acute hypoxic hyperventilation is dampened by two mechanisms. First, the hypocapnic alkalosis induced by hyperventilation yields a decrease in the CO_2-H^+ stimulus which, via both a chemoreflex mechanism and a central action, may depress the activity of respiratory neuronal pools in the brain stem. Second, in addition to exciting respiratory neurons, the arterial chemoreceptor afferents excite the reticular activating system which, in turn, may have composite action on ventilation (Sect. 3.2.1.3 b). Third, the afferent discharge of arterial chemoreceptors may change under the control of efferent pathways; such a control, however, is poorly understood. In any event, these interactions have been found to be dependent upon the excitability of the respiratory centers; consequently, no experiments under anesthesia (which may alter this excitability) should be regarded as definitive unless similar results can be obtained in unanesthetized preparations. Only experiments on conscious subjects are considered below.

3.2.1.2 Comparative Overview

None of the mammalian or bird species so far studied lacks chemoreceptors and ventilatory chemoreflexes (Adams 1958; Dejours 1976). Yet the part played respectively by the carotid bodies and by the aortic bodies in the normal control of ventilation remains undetermined. Some species, such as rabbits and pigs, may have no functional aortic chemoreceptors (Verbrugghe et al. 1982). Others, such as rats and mice, may possess abdominal chemoreceptors (Howe et al. 1981). In birds, chemoreceptive tissue has been reported to be present in various locations scattered through the neck region (De Kock 1959); the carotid bodies, however, were found to be entirely responsible for the ventilatory response to hypoxia (Jones and Purves 1970; Bouverot et al. 1979).

Data on the lower vertebrates are scarce. Morphological characteristics suggest that chemoreceptors are present in the carotid region of reptiles and amphibians; in the latter group physiological and electrophysiological evidence has been provided (Ishii et al. 1966). In the carp and trout, Eclancher (1975) demonstrated that the ventilatory reactions to fast changes of oxygenation are like those observed in the higher vertebrates (see below). Presumably chemoreceptors exist in fish, but their location is still uncertain and it has not been possible to suppress the ventilatory response to hypoxia by section of afferent pathways (Randall and Jones 1973; Bamford 1974).

Neuroepithelial cells have been described within the walls of airways of humans and other mammals (Lauweryns et al. 1972), birds, reptiles, and amphibians (see Widdicombe 1981), as well as within the primary epithelium of fish gill (Dunel-Erb et al. 1982). Their histophysiological features make these cells closely related to carotid body cells; hence they may be oxygen-sensitive (Lauweryns and Cokelaere 1973). Their physiological role, however, remains obscure to date.

3.2.1.3 Magnitude

In man and other mammals, numerous methods have been used to study the ventilatory response to hypoxia, with P_{CO_2} either allowed to fall as in nature, or kept steady, or increased further by breathing of CO_2 (Severinghaus 1972; Weil and Zwillich 1976). Although contributing information concerning the control of breathing, these methods do not help in determining how much of the overall ventilatory drive is provided by the arterial chemoreceptors. In contrast, it is possible to do so by means of the O_2-test method (Dejours 1957, 1962, 1981), the rationale of which is briefly recalled as follows.

If, in a given condition, a chemoreflex hypoxic drive is present, a decrease of ventilation must occur a few seconds after the start of an abrupt inhalation of a high O_2 mixture: P_{O_2} at the gas exchange surface increases at the first breath; after a few seconds, the blood with higher P_{O_2} flowing from this surface perfuses the chemoreceptor organs, the activity of which then decreases; at the same time a decrease in ventilation must be observed. By inference from mammalian studies, the ventilatory reaction is not the result of a change of the mechanical properties of the ventilatory pump (Dejours 1962). Positive support for the chemoreflex interpretation of the reaction is provided by the demonstration that (1) section of the afferent fibers from the carotid and aortic bodies to the respiratory centers abolishes the ventilatory reflex in the dog (Bouverot et al. 1965) while the bilateral block of IX and X cranial nerves abolishes it in man (Guz et al. 1966), and (2) during an O_2-test, both the electrical activity afferent from the carotid body and the ventilation decrease in parallel time courses in the cat (Leitner et al. 1965).

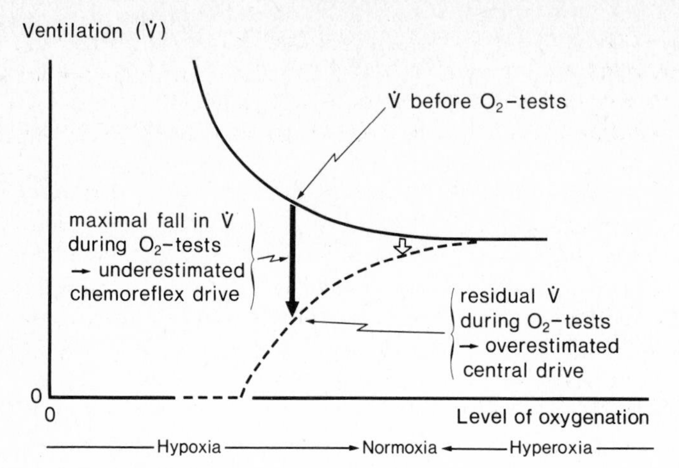

Fig. 3.6. Schematic representation of the ventilatory response to hypoxia and magnitude of the hypoxic chemoreflex drive. *Abscissa* level of oxygenation, as determined by the partial pressure of oxygen in the arterial blood, just before switching the subject to pure O_2 (O_2-test). *Ordinate* minute volume measured breath by breath. *Solid line* refers to the total ventilation in steady state, at the various levels of oxygenation. *Dashed line* refers to the minimal, or residual ventilation observed about 20–30 s after the onset of pure O_2 breathing (O_2-test) which raises arterial P_{O_2} above the threshold of stimulation of the arterial chemoreceptors, hence suppressing their activity. When the control level of oxygenation is below that threshold, the two curves diverge; the control total ventilation increases (*solid line*), and there is a trend toward central depression (*dashed line*), which reinforces as the level of oxygenation is made lower. Below a critical level of oxygenation, the central depression is such that the residual ventilation in the course of O_2 tests is zero. Note that (1) the magnitude of the hypoxic chemoreflex drive, as estimated from the vertical distance between the solid and dashed lines, is significant in normoxia (*white arrow*) and increases in hypoxia (*black arrow*), and (2) ventilation is entirely driven by afferent activity from the arterial chemoreceptors when the residual ventilation is zero. (Bouverot and Sébert 1979; Ungar and Bouverot 1980)

An estimate of the hypoxic chemoreflex drive which prevails at any level of oxygenation can be taken from the maximal decrease of ventilation (measured breath-by-breath) which is observed a few seconds after the switch to pure O_2 (vertical arrows in Fig. 3.6), before the secondary reactions to prolonged O_2 breathing (hypercapnia, acidosis, change in cerebral blood flow, etc.) come into play, and make the ventilation return toward higher values (see Fig. 1 in Ungar and Bouverot 1980). Within these few seconds, only the slight acidification of blood, due to increased oxygenation, cannot be avoided (Dejours 1962); since the related H^+-stimulus makes the residual minimum ventilation somewhat higher, this weighs against the arguments to be advanced below, thus strengthening the conclusions. Except when otherwise specified, the results that follow have been obtained at neutral ambient temperature, in conscious, resting subjects showing no sign of emotional constraint; neutral temperature is especially important for those animals which use their ventilatory system to dissipate heat.

a) Normoxia: In man near sea level, it is estimated that 10–15% of the ventilation is due to the O_2 chemoreflex drive (Dejours 1962). The fraction is higher in dogs (35%) than in man, and higher in rats (50%) than in dogs (Bouverot 1976); it proves to be about 30% in chickens, ducks, and pigeons (Bouverot and

Sébert 1979). The mechanisms underlying these species differences are unclear. On the other hand, how the chemoreflex drive of ventilation is divided between the carotid and aortic bodies remains to be determined in most species. Denervation of the carotid body, however, reduces this drive by half in dogs (Bouverot and Bureau 1975; Bouverot et al. 1981) and abolishes it in rabbits (Bouverot et al. 1973) and ducks (Bouverot et al. 1979).

b) Acute hypoxia: Since P_{O_2} is lowered in blood, ventilation in most cases increases immediately due to stimulation of arterial chemoreceptors. As shown in Fig. 3.6, the lower the level of oxygenation, the greater both the total ventilation (solid line) and the chemoreflex drive (vertical distance between solid and dashed curves). There are interspecific differences, however. For instance, at the altitude of 3.8–4 km, the hypoxic chemoreflex mechanism appears to drive 30–50% of the ventilation in man (Dejours et al. 1959, 1963), 50–60% in domestic birds (Bouverot and Sébert 1979), 90–100% in dogs or pigs (Ungar and Bouverot 1980; Verbrugghe et al. 1981). The magnitude of the chemoreflex drive is not known in most studies, but there appear to be interspecific differences in the hypoxic ventilatory threshold. This latter, for instance, is reported to be very low in the bar headed goose (Black and Tenney 1980, open triangles in Fig. 3.1). In these animals, it would be appropriate to know how much of the ventilation is driven by arterial chemoreceptors, above and below that threshold.

Below a critical level of oxygenation, the pulmonary ventilation appears to be entirely driven by the afferent activity from arterial chemoreceptors, at least in dogs, pigs, and domestic birds (above references). This is schematically illustrated in Fig. 3.6, in which the residual ventilation, when the arterial chemoreceptors are not operating (dashed line), falls to zero. At least three explanations can be proposed. First, the chemoreflex and centrogenic actions of the CO_2-H^+ stimulus decrease in proportion of the hypoxic hyperventilation, which induces hypocapnia and alkalosis. Second, whereas hypoxia excites ventilation through the chemoreflex mechanism, it inhibits ventilation centrally (Lahiri 1976; St. John and Wang 1977). Conceivably, therefore, the central drive of ventilation decreases due to the combined, central actions of hypoxia and of lowered CO_2-H^+ stimulus, whereas at the level of arterial chemoreceptors the stimulating action of hypoxia overrides the inhibiting influence of the lowered CO_2-H^+ stimulus. Third, an indirect cortical inhibitory influences may develop through suprapontine mechanisms (Tenney et al. 1971; Tenney and Ou 1977a).

With regard to this latter mechanism, discharging arterial chemoreceptors are known to excite not only the respiratory neurons via specific afferents, but also the reticular activating system, which, in turn, causes cortical arousal (Hugelin et al. 1959). The cortex, in turn, may exert descending inhibitory influences on the medullary respiratory centers. Might the frontal lobes contribute in some way, since stimulation of their orbital surface has long been known to produce a predominantly inhibitory effect on the ventilatory frequency? No unique contribution of the orbital gyrus could be documented by Tenney and St. John (1980). According to these authors, it remains conjectural whether a localized cortical function plays a significant role in the process of ventilatory adaptation to hypoxia.

c) Prolonged and life-long hypoxia: In lowlanders sojourning at high altitude, the magnitude of the hypoxic chemoreflex drive appears to be unaffected in the course of a few weeks exposure (see Dejours 1962). However, after a length of time still not precisely known, it appears to fall, although it never returns to sea-level value. Thus, in high altitude natives (born and raised at an elevation ranging from 3.5 to 5 km), the chemoreflex mechanism has been found to drive about 20% of the ventilation, which was less than in newcomers (Velásquez et al. 1968). Only in some patients with excessive polycythemia (chronic mountain sickness) and depressed ventilatory response to hypoxia, is the chemoreflex drive completely lost (Lefrançois et al. 1968; Lahiri 1968).

Comparable studies in animals are virtually lacking. The hypoxic chemoreflex drive has been reported to be present in highland native dogs (Lefrançois et al. 1968), but its magnitude in normal life cannot be assessed since the study was performed under general anesthesia. Also, little is known about the hypoxic chemoreflex drive in newborn at high altitude (see Dejours 1962; Lahiri et al. 1978).

3.2.1.4 Consequence of Its Loss

Integrity of the hypoxic chemoreflex drive of ventilation has been found essential in awake rabbits (Chalmers et al. 1967; Bouverot et al. 1973), dogs (Watt et al. 1943; Bouverot et al. 1965, 1973; Bouverot and Bureau 1975), ponies (Forster et al. 1976), goats (Forster et al. 1981; Lahiri et al. 1981), sheep (Lahiri et al. 1981), ducks (Bouverot et al. 1979), and in man (Wade et al. 1970), in determining the eupneic level of ventilation at low altitude as well as the normal ventilatory adaptation to hypoxia.

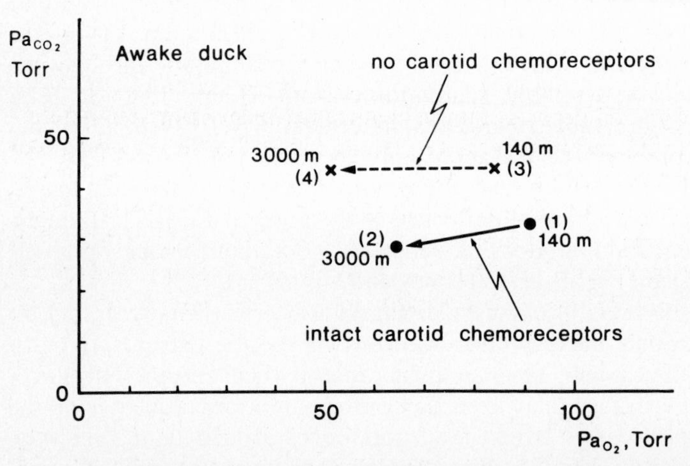

Fig. 3.7. Arterial blood P_{O_2} and P_{CO_2} in awake Pekin duck breathing air at neutral temperature at 140 m and 3000 m, before and after carotid body chemodenervation. The duck was studied (1) intact at 140 m, (2) intact after about 2 weeks at 3000 m; Pa_{CO_2} fell, concomitant with the hypoxic hyperventilation; (3) at 140 m, 2 weeks after bilateral surgical denervation of the carotid bodies, when arterial chemoreflexes were shown to be completely abolished; (4) again after about 2 weeks at 3000 m, when Pa_{CO_2} remained as at low altitude. The blood was sampled through an indwelling catheter implanted in the aorta via one brachial artery; the outer end of the catheter was coiled and taped to the dorsal side of the wing. (Bouverot et al. 1979)

The continuous role of the arterial chemoreceptors is well demonstrated by the effects of their *complete* surgical elimination. Near sea level, these effects consist of a permanent decrease of the effective ventilation [$\dot{V}$eff in Eq. (3.6)], with resulting decrease of P_{O_2} at the gas-exchange surface and in arterial blood, hypercapnia and acidosis. At high altitude, the process of ventilatory adaptation is no longer possible: due to the lack of hyperventilation, the chemodenervated animals remain hypercapnic and acidotic, and their arterial P_{O_2} is much lower than that of intact animals at identical elevation (Fig. 3.7); in fact, the highest altitude they can reach without danger to life is lower than for intact animals (Bouverot et al. 1973; Lahiri et al. 1981; Verbrugghe et al. 1981).

When, as in many studies, the only arterial chemoreceptors surgically denervated are those of the carotid bodies, the hyperventilatory process during sojourn at high altitude is only delayed and attenuated. The degree of attenuation varies among species, which has led to conflicting view (see Dempsey and Forster 1982). The most plausible explanation is a gradual resumption of the chemoreflex drive by previously dormant arterial chemoreceptors. Those in the aortic bodies are likely candidates (Bisgard et al. 1976; Smith and Mills 1980), but incomplete carotid body denervation, chemosensitivity of regenerated axons of the carotid sinus nerve (Mitchell et al. 1972), presence of ectopic chemoreceptor tissue, and other mechanisms, not yet known, cannot be excluded (Steinbrook et al. 1983). In any event, no definite conclusion regarding the process of ventilatory adaptation to hypoxia in supposedly chemodenervated subjects should be proposed without testing the existence or the actual suppression of the chemoreflex drive by an appropriate method, such as the O_2-test method (Sect. 3.2.1.3) in conscious state.

3.2.2 Central Mechanisms

Ventilation during hypoxia depends on the balance between stimulating and inhibiting influences to the respiratory centers. If, as stressed in the foregoing section, the arterial chemoreceptors are a dominant source of excitatory influence, the progressive increase in ventilation, which is observed in all the aerial vertebrates so far studied when going from acute to prolonged hypoxia (path 2→3 in Fig. 3.3 B), remains unexplained. Only a short account of this controversial problem will be given here, as Dempsey and Forster (1982) have recently reviewed the topic extensively.

3.2.2.1 Hyperventilatory Responsiveness

At the onset, it must be realized that the progressive return of P_{O_2} in the alveolar gas and arterial blood toward higher values, the drop of P_{CO_2} and alkalotic changes in arterial pH (path 2→3 in Fig. 3.3 B) are the consequence, not the cause, of the further increase in ventilation. When considered in isolation, and in the absence of information supporting an increase in chemoreceptor gain, these *decreasing* O_2, CO_2, and H^+ stimuli cannot account for the progressive increase in ventilation.

On the other hand, not only the magnitude, but also the nature of the acute and secondary ventilatory responses are different. Thus, the early increase in ventilation produced by acute hypoxia is immediately reversible when returning to normoxia, whereas the further, time-dependent increase is not: some hyperventilation, with its accompanying hypocapnia, continues after returning to normal oxygenation; only after hours or days will the ventilation and P_{CO_2} return to sea-level control values. It has also been repeatedly demonstrated (references in Dempsey and Forster 1982) that the synergism of hypoxia with other ventilatory stimuli is reinforced when hypoxia is prolonged. Thus, if P_{CO_2} is increased in the arterial blood by adding CO_2 to the inspired gas, it may be seen that the increment in ventilation per Torr increment in P_{CO_2} is greater in prolonged than in acute hypoxia. The same holds true for the increment in ventilation per increment of oxygen uptake under exercise. Hence, quoting directly from Dempsey and Forster (1982, p 302), the time-dependent changes in ventilation during hypoxia "appear to reflect a general hyperventilatory responsiveness." What are the underlying mechanisms?

3.2.2.2 Some Unsolved Problems

To explain the further increase in ventilation that occurs over several hours or days of hypoxia, it was early thought that the initial alkalotic changes in arterial pH inhibits a further hyperventilation until renal excretion of bicarbonate returns the arterial pH to normal (Rahn and Otis 1949). Another explanation was that similar acid-base changes occur in the cerebrospinal fluid, thus affecting the extra-cellular environment of the brain stem (Severinghaus et al. 1963). Today, both explanations are questioned because the time course of changes in the pH of either arterial blood or cerebrospinal fluid is not consistent with that of the changes in ventilation (Forster et al. 1975; Bureau and Bouverot 1975). Does the "apparent hyperventilatory responsiveness" during hypoxia then relate to some changes in the acid-base status in some other compartment of brain, as interstitial or/and intracellular fluids? Does it relate to lactic acidosis from glycolysis in brain cells elicited by hypoxemia and/or by alkalosis? Is it mediated by local ion shifts due to changes in cerebral blood flow? Is there a change in the excitability of the medullary respiratory neurons, and/or a change in chemoreceptor excitability? Do neurophysiological changes come into play, due to changes in the metabolism of central neurotransmitters (Chap. 6.3.1)? An analysis of current knowledge about these controversial questions has been provided by Dempsey and Forster (1982); there is no clear answer yet.

3.2.2.3 A Speculative Overview

On the basis of the observations made on conscious mammals and birds schematically illustrated in Fig. 3.6, one may postulate that alleviation of the central hypoxic depression may be responsible for the "hyperventilatory responsiveness" that develops progressively in the course of hypoxia.

Any point of the solid line in Fig. 3.6 refers to the (total) ventilation observed at the corresponding level of oxygenation read on the abscissa; this ventilation results from the balance between all the stimulating and inhibiting influences to the respiratory centers which prevail at that level of oxygenation. On the other hand, any point on the dashed line refers to the part of that ventilation that is driven by those influences which do not originate from arterial chemoreceptors. Let us call these influences the "central" drive.

Figure 3.6 shows that (1) even a change from hyperoxia to normoxia results in depression of the central drive, and (2) the lower the level of oxygenation, the worse the central depression. This is true even when correcting the hypocapnia induced by hypoxic hyperventilation, at least in the dog (Ungar and Bouverot 1980). These observations in conscious animals are in close agreement with electrophysiological data (Koepchen et al. 1976; St. John and Wang 1977). They are also in keeping with results indicating that hypoxia has deleterious central effects (Gemmill and Reeves 1933; Cherniack et al. 1970/71).

Assume now that partial alleviation of the central depression occurs during prolonged hypoxia, for instance as a result of efficient hematologic, vascular, and tissue changes, and consequently better coupling between O_2 transport by blood, metabolism, and neural function (Betz 1972; Kuschinsky and Wahl 1978). Then, the "central drive" will increase. In Fig. 3.6, this will yield an upward shift of the dashed line. The same will be true for the solid line, hence for the total ventilation, provided the magnitude of the chemoreflex drive remains unaffected.

If alleviation of the central hypoxic depression may explain the secondary increase of ventilation in the course of hypoxic exposure, the underlying mechanisms remain to be unraveled. Arterial chemoreceptor input, however, appears to be essential, not only in providing an important drive for ventilation, thereby favoring better oxygenation, but also in acting on the reticular activating system, thereby influencing various cortical and diencephalic functions (Gautier and Bonora 1980, 1982). The functional integrity of systems rostral to the pons also appears to be critical (Tenney and Ou 1977 b).

3.3 Efficiency

3.3.1 At Rest

It costs energy to propel air into and out of lungs, or water over gills.

In aquatic animals, the problem of cost of breathing may be especially important due to the high density and viscosity of water as compared to air. Unfortunately, many technical difficulties make the cost of breathing very difficult to assess (Ballintijn 1972; Holeton 1980); so that to date there are no reliable data. If the counter-current system allows the aquatic animals to remove O_2 from the ambient medium more efficiently than aerial vertebrates, it has been seen (Sect. 3.1.1) that, even in normoxia, the specific ventilation, $\dot{V}/\dot{M}_{O_2}$, is much greater in the former than in the latter. Furthermore, under hypoxia the O_2 extraction efficiency of aquatic animals decreases, and $\dot{V}/\dot{M}_{O_2}$ increases in greater proportion than the proportionate decrease in environmental oxygen [Eq. (3.4); Fig. 3.1]; theoretically, therefore, the work of pumping water should rise sharply. However, the gill resistance appears to be markedly reduced at high ventilation rate (see Ballintijn 1972), thereby reducing the energy expenditure of the respiratory muscles. On the other hand, in the course of prolonged hypoxia, subsequent readjustments in the water-to-blood diffusion barrier may restore, at least partially, a high O_2 extraction efficiency, hence reducing the ventilatory needs (arrow between closed squares in Fig. 3.1) and cost of breathing (Holeton 1980).

In aerial animals made hypoxic, the gain from hyperventilation is probably obtained at very low cost, as compared with water-breathers. In man, it has been estimated that the increased cost of breathing does not exceed 3% of the increased O_2 transport due to hyperventilation (Torrance et al. 1970/71). Much remains to be learned, however, since the mechanics of breathing can be affected by changes in lung volume, lung interstitial fluid, pulmonary hypertension, and complex changes of airway resistance under the opposite actions of hypoxia, hypocapnia, and lowered gas density at high altitudes (Cerretelli 1980).

Enhanced gas exchange across the medium/blood interface may be a saving mechanism, allowing aerial vertebrates to insure adequate blood loading of O_2 at minimal cost of breathing. Thus the bar headed goose does not hyperventilate, although retaining normal oxygen uptake, until very severe hypoxia (Fig. 3.1, open triangles). In this bird, a high affinity of hemoglobin for O_2 (Black and Tenney 1980, Chap. 4), may be more important than the cross-current system (Scheid 1982, Chap. 2.2.1b).

3.3.2 During Exercise

Data obtained in human beings (Dejours et al. 1963; Cerretelli 1980) indicate that for any level of oxygen uptake [$\dot{M}_{O_2}$ in Eq. (3.2)] the pulmonary ventilation is always higher at high altitude than at sea level. Not only the slope of the $\dot{V}$ vs. $\dot{M}_{O_2}$ relationship is greater in acute hypoxia than in normoxia, but it is greater in prolonged stages than in acute stages of hypoxic exposure. This means that, due to the positive interaction of the stimuli which come into play, the O_2-driving pressure at the gas exchange surface improves with time [E_{O_2} decreasing in Eq. (3.2), and $Pout_{O_2}$ increasing in Eq. (3.5)]. Integrity of the hypoxic chemoreflex drive seems to be critical (Bouverot et al. 1981). Enhanced alveolar-arterial O_2 transfer (Dempsey et al. 1971) allows highlanders to hyperventilate less than sea-level dwellers at the same elevation and identical O_2 consumption. Thus the Sherpas exhibit maximal exercise ventilation attenuated by about 20–30% as compared with that of acclimatized lowlanders (Pugh et al. 1964; Lahiri et al. 1972).

3.3.3 During Sleep

Depression of hypoxic hyperventilation, and therefore worsening of hypoxemia, are reported to occur during sleep at high altitude. Thus, Reed and Kellogg (1960) found that the pulmonary ventilation and the arterial oxygen saturation decreased in sea-level dwellers when they fell asleep at the altitude of 4.34 km: these subjects became as hypoxic as if they had abruptly ascended to about 5 km. On the other hand, it has long been known that sleep disturbance is common during the first day or two following ascent to high altitude (see Bert 1978). Insomnia, frequent awakenings, sensation of not having breathed, and headache which usually worsens in the morning upon rising are often experienced by persons who ascent to high altitude too quickly. Electroencephalographic studies show an increased frequency of arousals, associated with marked periodic breathing, i.e., a recurrent pattern of hyperpneic and apneic episodes (Reite et al. 1975).

The cause of periodic breathing and its link with worsening hypoxemia remain unclear, in spite of much investigative effort prompted by clinical disorders such as adult sleep apnea syndrome and sudden infant death syndrome, observed even at sea level (references in Remmers 1981; Strohl 1982). Limited data suggest that sleep depresses the central action of the CO_2-H^+ stimulus, whereas it has little if any effect on the magnitude of hypoxic chemoreflex drive (Reed and Kellogg 1960). Conceivably, therefore, a decrease in central drive may be responsible for the depression of hyperventilation and worsening hypoxemia during sleep.

Weil et al. (1978) found that periodic breathing is neither decreased by oxygen administration nor exaggerated by hypoxia, whereas it is relieved by CO_2 administration. This observation suggests that hypocapnia rather than hypoxia is significantly related to periodic breathing. This view is supported by other observations (Weil et al. 1978; Sutton et al. 1980), indicating that the amount of periodic breathing during sleep is reduced, the average arterial oxygen saturation improved, and the number of arousals decreased when acetazolamide is administered before ascent to high altitude. This carbonic anhydrase inhibitor promotes a bicarbonate diuresis, and therefore a metabolic acidosis which largely prevents the alkalosis induced by hypoxic hyperventilation (Cain and Dunn 1966). On the other hand, Lahiri et al. (1983) investigated these questions at the Base Camp Laboratory during the American Medical Research Expedition to Everest, 1981. They concluded that periodic breathing during sleep at high altitude results from the interplay between multiple factors. The primary determinants, however, appear to be the arterial chemoreflex controlling ventilation and the associated respiratory alkalosis. None of the Sherpa highlanders with low ventilatory sensitivity to hypoxia showed any sustained periodic breathing with apnea.

The mechanism responsible for arousals remains elusive. Arterial chemoreceptors are likely to be involved (Guazzi and Freis 1969; Bowes et al. 1981), due to the cortical activation that they provoke through the reticular activating system (Hugelin et al. 1959). Neubauer et al. (1981) judged this view incompatible with their finding that arousals occur at the same blood O_2 saturation in intact and carotid chemodenervated cats, as well as under carbon monoxide inhalation; however, aortic chemoreceptors are stimulated by hypoxia, and by even a small increase in carboxyhemoglobinemia (Lahiri 1980).

3.3.4 During Pregnancy

Although pregnancy is known to increase ventilation (and to decrease hemoglobin concentration) at low altitude, existing observations on maternal O_2 transport at high altitude are virtually lacking. Studying 44 residents of Leadville (Colorado; altitude 3.1 km), Moore et al. (1982a, b) found that birth weight was related to absolute maternal ventilation as well as to changes in ventilation during pregnancy. Smaller babies were delivered by women whose minute volume was lower and in whom the ventilation failed to increase from early to late gestation. However, these "hypoventilating" women did not have lower arterial O_2 saturations or higher arterial P_{CO_2}, nor was arterial O_2 saturation related to infant birth weight. Obviously further studies are needed.

3.4 Concluding Remarks

Hyperventilation is an important factor of adaptation to altitude hypoxia. Its primary effect is to raise the O_2 partial pressure at the medium/blood interface, and therefore the driving pressure for blood O_2 loading and subsequent transfer of oxygen to cells. Hypocapnia and alkalosis are induced factors. Hypocapnia persists as long as hyperventilation is present, whereas alkalosis is corrected, at least partially, essentially by bicarbonate excretion from blood.

Judging from the *proportion* of oxygen that they can extract from inhaled medium, water-breathers appear to be best equipped to cope with low environmental availability of O_2; among the air-breathers so far studied, only high-altitude native birds, such as the bar headed goose, bear comparison with them. Yet, the *amount* of oxygen that water-breathers are able to take up from a unit volume of medium is very small, due to the very low solubility and concentration of O_2 in waters; consequently, at identical O_2 partial pressure in the inspired medium, ventilatory requirements are much greater in water-breathers than in air-breathers.

How hypoxic hyperventilation is mediated in water-breathers is not precisely known. In contrast, all the aerial vertebrates studied so far are dependent on the reflex drive that originates in arterial chemoreceptors. How the magnitude of this drive varies among species at identical levels of oxygenation is neither fully documented nor understood. The same holds for the mechanisms responsible for the time-dependent changes of hyperventilation observed among individuals at a given elevation. There seems to be a central neural depression even in normoxia, reinforced in the early phase of hypoxia as it is overridden by the chemoreflex drive. The fundamental mechanisms for this central depression, and whether it is alleviated during prolonged hypoxia, are unsolved problems.

Chapter 4

Circulatory Adaptations

As already indicated in Fig. 2.2, the cascade in O_2 partial pressure from arterial to venous blood decreases under altitude hypoxia. This means that an adaptive process must occur in the transport of O_2 by blood, that prevents the O_2 pressure in the blood flowing through tissue capillaries from falling greatly during hypoxia.

4.1 An Overview

Transport of O_2 by blood has a dual function: (1) it must keep the flux of O_2 sufficient to match the metabolic demand, while (2) it must provide an adequate pressure head for O_2 diffusion through tissues.

The first function implies that, in a steady state, the following Fick's equation for the whole organism is satisfied

$$\dot{M}_{O_2} = \dot{V}b\,(Ca_{O_2} - C\bar{v}_{O_2}), \qquad (4.1)$$

where $\dot{M}_{O_2}$ is the oxygen consumption, $\dot{V}b$ is the total (cardiac) blood flow, Ca_{O_2} and $C\bar{v}_{O_2}$ are the O_2 concentrations in arterial and mixed venous blood. Equation (4.1) is equivalent to the following

$$\dot{M}_{O_2} = \dot{V}b \cdot \beta b_{O_2} \cdot (Pa_{O_2} - P\bar{v}_{O_2}), \qquad (4.2)$$

in which $Pa_{O_2} - P\bar{v}_{O_2}$ is the arterial-venous P_{O_2} difference, and βb_{O_2} is the ratio $(Ca_{O_2} - C\bar{v}_{O_2})/(Pa_{O_2} - P\bar{v}_{O_2})$. Called the blood O_2 capacitance coefficient (Dejours et al. 1970), βb_{O_2} is the slope of the line joining the arterial point to the venous point on a C_{O_2} vs. P_{O_2} plot (Fig. 4.1 A). In contrast with the O_2 capacitance coefficients in air or water, which are constant at given temperature, βb_{O_2} varies with the level of oxygenation owing to the nonliner shape of the C_{O_2}–P_{O_2} relationship in blood. The product $\dot{V}b \cdot \beta b_{O_2}$ is the O_2 circulatory conductance.

The second function mentioned above becomes evident when rewriting Eq. (4.2) as follows

$$P\bar{v}_{O_2} = Pa_{O_2} - \frac{1}{\beta b_{O_2} \cdot \dot{V}b/\dot{M}_{O_2}} \qquad (4.3)$$

and viewing $P\bar{v}_{O_2}$ as the lowest O_2 pressure head at the vascular supply source for tissue oxygenation. In Eq. (4.3), the ratio $\dot{V}b/\dot{M}_{O_2}$ designates the volume of blood

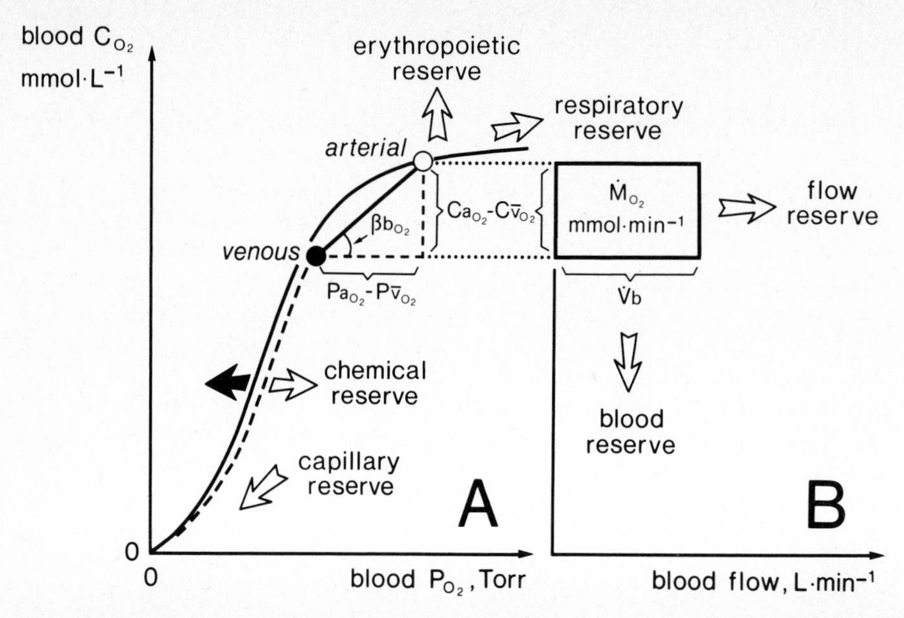

Fig. 4.1. Schematic representation of the O_2 transport by blood, and factors influencing the O_2 partial pressure in peripheral capillary blood. (After Metcalfe and Dhindsa 1970). *Ordinate* in *A* and *B:* O_2 concentration in whole blood (C_{O_2}). *Abscissae:* in *A* partial pressure of O_2 in blood (P_{O_2}); in *B* blood flow ($\dot{V}b$). *A:* blood O_2 equilibrium curves at the physicochemical conditions (temperature, P_{CO_2} and pH) prevailing in arterial blood (*a, solid curve, open circle)* and venous blood (*v, dashed curve, closed circle*); the slope of the line joining the arterial and venous points indicates the blood O_2 capacitance coefficient [βb_{O_2} in Eq. (4.2)]. *B*: the rectangle area is proportional to oxygen consumption [$\dot{M}_{O_2}$ in Eqs. (4.1) and (4.2)]. Labeled arrows refer to the factors that affect blood O_2 transport. Explanations in text

that must be passed over the gas exchange surfaces in tissues to effect delivery of a unit quantity of oxygen. It is termed the specific blood O_2 convection, or the blood convection requirement (Dejours et al. 1970). The product $\beta b_{O_2} \cdot \dot{V}b/\dot{M}_{O_2}$ is the specific blood O_2 conductance. Since partial pressure of O_2 in arterial blood, Pa_{O_2}, essentially results from ventilation and O_2 equilibration at the external medium/blood interface, Eq. (4.3) states that the greater the specific blood O_2 conductance, the more closely $P\bar{v}_{O_2}$ approaches Pa_{O_2}, and therefore the higher is the O_2 pressure head for subsequent O_2 transfer to tissues.

Derived from Metcalfe and Dhindsa (1970), Fig. 4.1 shows the close mutual involvement of the factors in Eqs. (4.1), (4.2), and (4.3); arrows visualize possible adaptive changes in the O_2 transport by blood. It is clear that the blood O_2 equilibrium curve is of central importance.

The above considerations are not limited to the whole body. The arterial blood is distributed to various tissues and organs in parallel, each one with its own oxygen needs, its own blood flow rate, and its own arteriovenous O_2 (and CO_2) differences of concentrations and of partial pressures. There are, therefore, as many O_2-equilibrium curves and βb_{O_2} values as tissues and organs, which may differ from the average values of the whole body; we will return to this point later on.

4.2 Blood O_2 Capacitance

Probably the most important compensatory mechanism in the struggle for a high venous P_{O_2} resides in the S shape of the blood O_2 equilibrium curve as such (Kreuzer 1966). Figure 4.1 shows clearly that, when going from normoxia to hypoxia, the representative points for arterial and venous blood shift from the flat upper part of the curve down along the straight steepest part. Then, the slope of the line joining the arterial and venous points, i.e., the capacitance coefficient of blood for O_2 [βb_{O_2} in Eqs. (4.2) and (4.3)], increases considerably (by a factor up to 5 or 6). This provides an automatic adjustment of blood O_2 conductance, thereby preventing the venous P_{O_2} from falling excessively under hypoxia, while keeping the $Ca_{O_2}-C\bar{v}_{O_2}$ difference and blood flow constant, and hence the net flux of O_2 molecules [$\dot{M}_{O_2}$ in Eq. (4.1)]. The process is operative whatever the duration of hypoxia, and contributes most significantly in preserving capillary P_{O_2}. Its efficacy, however, decreases slightly when the venous point is dropped down to the lowest curvilinear part of the O_2-equilibrium curve; this occurs under very severe hypoxia, and/or when the arterial-venous O_2 concentration difference widens to near maximum, as during an intense metabolic demand. In such conditions, the tissular "capillary reserve" (Fig. 4.1 A) is critical in allowing for better diffusion and utilization of O_2.

The S shape of the O_2-equilibrium curve, and therefore the blood O_2 capacitance, can change if the hemoglobin concentration increases ("erythropoietic reserve" in Fig. 4.1 A), or/and if the hemoglobin-oxygen affinity increases (black arrow labelled "chemical reserve") or decreases (white arrow), and/or if the heme–heme interaction is altered (not shown in Fig. 4.1; the stronger this interaction, the steeper the slope of the blood O_2 equilibrium curve). These factors are considered below.

4.2.1 Hemoglobin Concentration

An increase in the "erythropoietic reserve" (upper arrow in Fig. 4.1) has long been a consistent finding in humans and other mammals sojourning at high altitudes (Lenfant 1973). Along with polycythemia, the process increases both the concentration of hemoglobin and O_2, and makes the blood O_2 equilibrium curve steeper. Accordingly, assuming that the arterial P_{O_2}, blood flow, and oxygen consumption remain unchanged, the venous P_{O_2} becomes higher (the tissue oxygenation improves) because the blood O_2 capacitance [βb_{O_2} in Eq. (4.3)] is raised (Turek et al. 1973 a). There are observations, however, that make the adaptive significance of this process debatable.

4.2.1.1 Major Features

Viault (1890) was the first to report a prompt increase in the number of his own red cells after traveling by railroad from Lima at sea level to the Andean town of Morococha at the altitude of 4.54 km. Since that time an augmented hemato-

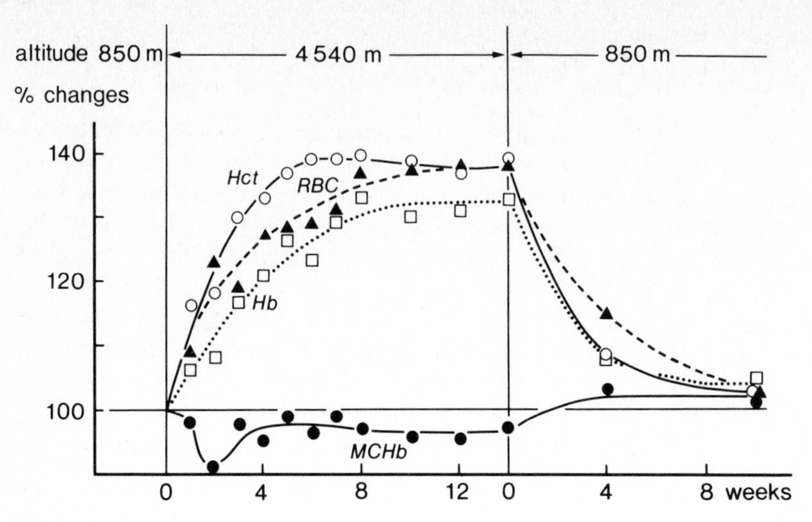

Fig. 4.2. Time course of some hematologic changes in man exposed for a few weeks at 4540 m. (Bododzhanov, in Altman and Dittmer 1974, p 1881). The control values before exposure to high altitude were the following: hematocrit, *Hct* 39%; erythrocyte count, *RBC* 4.58 millions μL^{-1}; total hemoglobin concentration in blood, *Hb* 132 $g \cdot L^{-1}$; mean hemoglobin concentration per erythrocyte; *MCHb* 28.8 pg $cell^{-1}$

poietic activity under hypoxia has been repeatedly confirmed (see Hurtado 1932, 1964; Hurtado et al. 1945; Bullard 1972; Lenfant 1973; Pace 1974; Monge and Whittembury 1976). There are, however, intra- and interspecific differences, not fully understood, which most probably depend upon many variables such as intensity and duration of hypoxia, concomitant cold stress (Lechner et al. 1981), stage of development at the time of exposure (Bartels et al. 1979; Smith et al. 1979), sex, nutritional, seasonal, and genetic factors.

Figure 4.2 shows a typical example in human beings translocated from low (0.85 km) to high altitude (4.54 km). Red cell count (RBC), hemoglobin concentration in blood (Hb), and hematocrit (Hct) increase roughly in parallel, while the mean hemoglobin concentration per red cell (MCHb), if anything, decreases (Hurtado 1932) due to either reduced amounts of intracellular hemoglobin, or swelling of the cells, or both. The process is slow and requires weeks or months to reach full expression. It also reverses slowly after return to low altitude. The magnitude of the erythropoietic response has been reported to increase linearly with elevation up to about 4 km, and more abruptly above this altitude; beyond a limit, close to 6 km for man, a decrease in the response begins (Hurtado et al. 1945).

Table 4.1 provides some comparative data. Reptiles are excluded because the data are contradictory. Some evidence indicates that, when increased hematocrit occurs in mountain lizards, the change correlates with cold and not with hypoxia (Bennett and Dawson 1976; Wood and Lenfant 1976; Hillyard 1981). In reptiles, the variable but usually high capacity for anaerobic activity is thought to minimize the need for hematological adaptations. In other ectotherms, such as amphibians and fishes, small but significant increases in hemoglobin concentration

Table 4.1. Hematocrit (Hct), red blood cell count (RBC) and hemoglobin concentration (Hb) in humans and animals at various altitudes or levels of oxygenation

Species	Condition	Hct (%)	RBC $10^6 \cdot \mu L^{-1}$	Hb $g \cdot L^{-1}$	Ref.
	Altitude m				
Humans					
Peruvians	0	46.6	5.1	156	1
–	4540 native	59.5	6.4	201	
Chileans	450	38.1		145	2
–	3200 native	45.6		168	
–	4460	54.3		184	
Italians	120	47.7	4.7	155	3
–	3500 accl.	53.3	5.4	176	
Sherpas	3850 native	50.0	4.7	170	
Camelids					
Alpaca	4200 native	35.2	14.4	138	4
Llama	– –	38.3	13.7	151	
Vicuña	4300 –	36.1	13.1	135	
Birds					
Pekin duck	0	45.4	3.2	155	5
–	5640 accl.	55.9	3.9	240	
Bar headed goose	0	47.8	3.0	139	5
–	5640 accl.	43.9	2.7	129	
Ectotherms					
Frog 10 °C	335 accl.	23.6	0.56	73	6
– 13 °C	3750 native	27.9	0.73	81	
	P_{O_2}, Torr				
Trout 20 °C	150	28.2		64	7
– –	80 accl.	28.5		65	
– –	50 –	35.3		72	
Carp 20 °C	120–150	26.4	0.95	74	8
– –	30 accl.	29.0	1.02	75	

References: 1. Hurtado (1964); 2. Clench et al. (1982); 3. Samaja et al. (1979); 4. Reynafarje (1966); 5. Black and Tenney (1980); 6. Hutchison et al. (1976); 7. Tetens and Lykkeboe (1981); 8. Weber and Lykkeboe (1978)

and hematocrit occur in response to hypoxia (Table 4.1). Although hematological changes have been repeatedly demonstrated in all homeotherms tested at high altitudes, no increased erythropoietic activity seems to exist in members of highland species (Bullard 1972; Monge and Whittenbury 1976). Table 4.1 shows that hematocrit and hemoglobin concentration are not elevated in the Andean camelids (alpaca, llama, and vicuña) and Himalayan bar headed geese, which successfully thrive at high altitude. A noteworthy feature is the large number of red cells in the Andean camelids; Reynafarje (1966) stressed that the very small size and elliptical shape of these cells might be of definite importance in increasing the surface area for gas exchange, thereby favoring red-cell oxygenation in these camelids. These characteristics, however, are also present in the low-altitude desert camelids in Africa, and therefore are not specific adaptations to high altitude (Monge and Whittembury 1976). On the other hand, guinea pigs native to high altitude (4.11 km) have higher hematocrits (46–47%) than sea-level controls (39%), but these increased altitude values are almost the same as the sea-level values of man and rat (Turek et al. 1980). Chronic exposure of immature animals to hypoxia may involve a more diverse adaptive response than that seen in adults; the characteristics of the red cell itself may change with respect to its size and shape, its Hb content, and its stage of maturity (Lechner et al. 1980).

As for human beings native to highlands, Table 4.1 indicates that hemoglobin concentration and hematocrit have been found to be lower in Sherpas than in Andean natives or acclimatized lowlanders, even though they are increased in comparison with sea-level values. Adams and Strang (1975) stressed that healthy persons of Tibetan ancestry living at 4 km altitude in the Nepal Himalayas have hematological values similar to that expected at 2.3 km in the Andes. Considerable variations have, however, been observed in Andean natives. Faced with such a variability, Hurtado (1932) concluded that "an increase in the erythrocytes of the circulating blood, although the common finding, is not an essential adaptation process for normal life at high altitudes" (p 503).

4.2.1.2 Efficiency

Whether polycythemia represents a favorable adaptation to hypoxia has been debated. The response is costly in that it increases metabolic activities, and may be outweighed by increased resistance to blood flow, since increasing the red-cell mass will increase blood viscosity (Crowell and Smith 1967).

Studying the effects of normovolemic anemia and polycythemia in dogs, Richardson and Guyton (1959) adjusted the hematocrit experimentally from 20 to 68%. They found that the blood O_2 transport [the product $\dot{V}b \cdot Ca_{O_2}$ or $\dot{V}b \cdot \beta_{O_2} \cdot Pa_{O_2}$; Eqs. (4.1) and (4.2)] was maximal at the normal hematocrit of 40%. Crowell and Smith (1967) confirmed this finding by both in vivo and in vitro studies. Thus, an increase in hematocrit beyond a certain value may be actually deleterious to oxygen transport. However, the "optimal" hematocrit is not precisely known. It may be shifted toward 50–60% for the delivery of O_2 to contracting muscles (Gaehtgens et al. 1979; Buick et al. 1980), but to much lower values for the maintenance of coronary blood flow and cardiac output (Kuramoto et al. 1980). Furthermore, hypoxia (Pa_{O_2} = 41 Torr) has been reported as magnifying the oxygen transport deficit induced by polycythemia (McGrath and Weil 1978).

A peripheral impairment to O_2 uptake during exercise, perhaps due to increased blood viscosity, may explain the failure of subjects acclimatized to 5.4 km and above to reach, at high altitude, the sea-level $\dot{M}_{O_2}^{max}$ while breathing pure O_2 in spite of a 40% increase of Hb concentration (Cerretelli 1976). On the other hand, a reduced hematocrit by removal of red cells from climbers may decrease $\dot{M}_{O_2}^{max}$ in spite of an increase in the maximal cardiac output (Horstmann et al. 1980).

No definitive statement of the efficiency of the increased red cell mass from hypoxia acclimation is possible at this time (see also Sect. 4.2.1.4).

4.2.1.3 Mechanism

Conceivably the increased red cell count and hemoglobin concentration during hypoxia may result from either a decrease in plasma volume (hemoconcentration) or an increase in the red cell mass. The latter may proceed from either a release of erythrocytes from reservoirs, such as the spleen, or from increased erythropoietic activity. The available information, essentially from mammalian studies, suggests that all these response patterns may come into play. Their respective contributions to polycythemia, however, are not precisely known, and will depend upon many factors such as species, severity of hypoxia, sex, age, body hydration, nutrient availability (especially iron), as well as associated stresses of exercise, cold exposure, emotional excitement, etc. Clearly, the factors triggering polycythemia remain elusive.

Hemoconcentration due to redistribution of water from the extracellular to intracellular compartment has been reported to occur in man within the first week of exposure to an altitude of 4.3 km (Surks et al. 1966). Since this study, there have been conflicting viewpoints about the body's handling of water and its underlying mechanisms (see Keynes et al. 1982; Hackett et al. 1981; Jones et al. 1981; Miles et al. 1981). Methodological biases may have influenced the results (Turek et al. 1978 a).

Contraction of the spleen resulting in the mobilization of additional red cells, thereby increasing O_2-carrying capacity of the blood, is known to occur in response to adrenergic stimulation and to severe hypoxia of short duration (Barcroft 1925; Kramer and Luft 1951; Cook and Alafi 1956).

With continued hypoxic exposure, there are absolute increases in the red cell number and total blood volume (Merino 1950). The daily red cell production increases, judging from the increased incorporation of radioactive iron, but the life span of red cells seems unaffected (Reynafarje 1964, 1966). Depending on the animal species, age and the severity of hypoxic stress, the kidney, spleen, liver, and bone marrow play key roles in the polycythemic response (Ou et al. 1980). The associated hyperplasia of the erythroid elements is thought to be triggered by the hormone erythropoietin (see Adamson and Finch 1975). The availability of oxygen to the tissues, particularly renal, concerned with induction of erythropoietic activity is presumably the major regulator of erythropoiesis, but the microenvironment of erythropoid cells, nutritional, hormonal, and neural factors also appear to be critical (Fisher 1977). The information concerned with the regulation of hemoglobin synthesis during the development of the red cells has been reviewed by Nienhuis and Benz (1977), but questions asked 20 years ago (Hurtado 1964) remains unanswered: Why does the erythropoietic activity develop only a few days after birth, indicating that the hypoxic stimulus is either not present or ineffective during uterine life? Why, in adults, does the plasmatic concentration of erythropoietin decline to nondetectable values within 2 days after ascent to high altitude, whereas the red cell count and hemoglobin concentration remain elevated throughout the stay at high altitude? How, then, is the increased erythropoietic activity triggered?

A role of arterial chemoreceptors in the control of erythropoiesis has been advocated (references in Paulo et al. 1973; Fisher 1977). It is true that denervation of carotid body enhances the erythropoietic response to hypoxia. Most probably this phenomenon is related to the observation that hypoxemia at a given altitude is more pronounced in the denervated animal (Fig. 3.7), due to the concurrent abolition or reduction of the ventilatory response to hypoxia after chemodenervation (Chap. 3.2.1.4).

4.2.1.4 Hyperexis

The term "hyperexis" (from a Greek word meaning "excess response") was proposed by Richards (1960) to designate an apparent homeostatic effort which, in correcting one perturbation, produces others much worse. The following quotation of a few lines which refer to the polycythemia observed in chronic pulmonary disease is appropriate: "Arterial anoxia[3] stimulates the bone marrow, and more red blood cells are formed, just as in the favorable response at high altitudes. But then things begin to go wrong. The increased red cell mass produces increased blood volume, which overfills the heart; and this organ, already under strain working against a restricted pulmonary vascular bed, goes into congestive failure, which still further increases blood volume. At the same time the added pulmonary vascular engorgement further aggravates the anoxia, thus setting up a second vicious cycle." Similar pathological "overshoots" are observed in some high altitude natives. Thus, the hematocrit and hemoglobin concentration may reach values up to 83% and $270\,g \cdot L^{-1}$, respectively, in some individuals living at altitudes above 3 km in the Andes. Such an excessive erythropoietic activity is part, if not the cause of the many symptoms of the "*chronic* mountain sickness," termed "Monge's disease" to commemorate the Peruvian physician who first described the syndrome (see Monge 1943). The etiology of Monge's disease is not precisely known; excessive polycythemia, however, relates to insensitivity of arterial chemoreceptors and alveolar hypoventilation, which increases hypoxemia (see Chap. 3.2.1.3c; Heath and Williams 1981). When those permanent residents who have marked polycythemia are bled over a period of 2 or 3 weeks to reduce their hematocrit to normal sea-level values, their exercise tolerance is sometimes improved (Winslow et al. 1979).

4.2.2 Blood O_2 Affinity

The foregoing section was concerned with the *quantity* of hemoglobin. We will now consider the *quality* of the pigment, or more precisely the factors that, under altitude hypoxia, may change the position and shape of the blood O_2 equilibrium curve (Fig. 4.1; arrows labeled "chemical reserve").

The relative position of the blood O_2 equilibrium curve (OEC) has long been viewed as a possible factor in modifying O_2 transport, and thus tolerance to hypoxemia. However, there is at present no general agreement about the physiological significance of the changes in OEC either produced experimentally or observed under normal high-altitude exposure.

3 "Hypoxia" would be the correct term

4.2.2.1 Basic Considerations

Several reviews (Bauer 1974; Shappell and Lenfant 1975; Bartels and Baumann 1977; Wood and Lenfant 1979; Duhm 1980), and some papers in a recent issue of the American Zoologist (1980, Vol 20, No 1) provide information on the structure–function relationships of hemoglobin and the basic interpretation of the physiological role of the OEC in adaptation to hypoxia. It is important to emphasize that the properties of hemoglobin in vivo such as O_2 affinity, heme–heme interaction, buffering capacity, Bohr effect, temperature sensitivity, etc. are complex functions of the molecular structure of the polypeptide subunits and their interaction with each other and with intracellular ligands (carbon dioxide, protons, chloride, organic phosphates) some of which, in turn, are regulated by erythrocyte metabolism.

Due to the metabolic control of hemoglobin function provided by erythrocytes, it must be realized that information regarding the *physiological implications* of the position of the OEC is valid only if it is obtained under conditions that prevail in vivo. As yet, this has not been the case in many studies, a defect that renders their interpretation difficult or uncertain.

4.2.2.2 Theoretical Studies

Figure 4.3 shows that (1) *under moderate hypoxia* (open symbols) a right-shifted OEC makes the line joining arterial and venous points steeper [βb_{O_2} increases in Eq. (4.3)] and raises $P\bar{v}_{O_2}$, the overall index of tissue oxygenation, (2) *under severe hypoxia* (closed symbols), the reversed situation prevails, both βb_{O_2} and $P\bar{v}_{O_2}$ being greater with a left-shifted OEC than with a right-shifted one. These possibilities exactly reflect the conclusion of Turek et al. (1973) from a theoretical study comparing man ("low" O_2 affinity) and rat ("high" O_2 affinity) in resting conditions. According to these authors, the critical value of arterial P_{O_2} below which a right-shifted OEC may prove detrimental would be around 35 Torr. Lahiri (1975) derived similar results; but he found that the critical Pa_{O_2} rises to 60 Torr when the arterial-venous O_2 concentration difference enlarges, as in the body as a whole during physical activity.

The above distinction between moderate and severe hypoxia is, however, questionable. Even at low altitude a left-shifted OEC may be adaptive (Bencowitz et al. 1982). If, as in exercising man, the kinetics of O_2 transfer across the external medium/blood barrier becomes a limiting step (diffusion limitation; Chap. 5.3.3), the arterial P_{O_2} will decrease (in Fig. 4.3 the highest open symbols would be shifted down along their respective curves to below 80 Torr). Then, an increased O_2 affinity will limit the decline of Pa_{O_2}; using Fig. 4.3 to illustrate this, the open triangle would shift down along OEC, but not as much as would the open circle. This important relationship can perhaps be better understood when focusing attention on the shape of the upper part of the O_2 equilibrium curves. When going from the venous point to the arterial point, while blood is loaded with oxygen during its passage through the pulmonary capillary, it is evident that, for any increment in O_2 concentration, a left-shifted OEC leads to a greater increase in P_{O_2}, in comparison with a right-shifted OEC. It is exactly what the computation shows (Bencowitz et al. 1982): *with diffusion limitation,* a left-shifted OEC not only

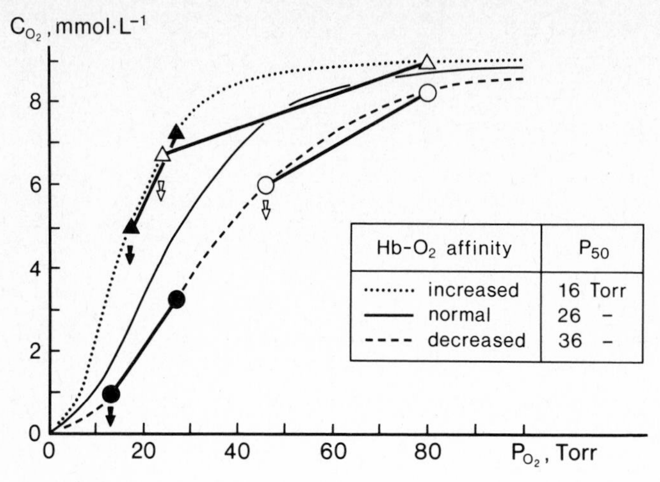

Fig. 4.3. Theoretical influence of a change in hemoglobin (Hb) O_2 affinity on the transport of oxygen by blood under moderate hypoxia (*open symbols*) and severe hypoxia (*closed symbols*). *Ordinate* O_2 concentration in whole blood (C_{O_2}); *abscissa* O_2 partial pressure in blood (P_{O_2}). *Symbols labeled with arrows* venous blood; *other symbols* arterial blood. *Sigmoid curves* in vitro blood O_2 equilibrium curves at normal, increased and decreased O_2 affinity, as precised in the inset; $P_{50}=O_2$ partial pressure at half saturation of the pigment with oxygen (at 37 °C, pH=7.4; $P_{CO_2}=40$ Torr). For clarity, the changes in Hb-O_2 affinity are presumed to maintain the values of the O_2 partial pressure in arterial blood (Pa_{O_2}), oxygen consumption ($\dot{M}_{O_2}$) and blood flow ($\dot{V}b$) unaffected [Eqs. (4.1)–(4.3)]; consequently, all arterio-venous O_2 concentration differences are identical. At a given level of hypoxia, note the effects of changing Hb-O_2 affinity on the value of the O_2 partial pressure in mixed venous blood (*arrows*), and on that of the blood O_2 capacitance coefficient (*slopes* of the line joining related arterial and mixed venous points). Other explanations in text

makes the arterial P_{O_2} higher, but also prevents both the venous P_{O_2} and maximal oxygen consumption from falling as much as they would with right-shifted curve. On the other hand, *with no diffusion limitation,* Bencowitz et al. (1982) confirmed the conclusion of Turek et al. (1973a) that, under moderate hypoxia ($Pa_{O_2}>45$ Torr), a left-shifted OEC will be unfavorable, because it will give the lowest values of venous P_{O_2} at all levels of activity. Accordingly, Bencowitz et al. (1982) concluded that "whether a leftward or rightward shift of the OEC is advantageous to gas exchange depends on the presence or absence of diffusion limitation."

An altered heme–heme interaction, or cooperativity, as described by the Hill number, might be of significant adaptive value in some circumstances (Neville 1977 a, b; Teisseire et al. 1979).

4.2.2.3 Experimental Data

Despite the fact that the position of the blood O_2 equilibrium curve during hypoxia has been intensively investigated for more than half a century, there is still no general agreement as to what changes are of adaptive value.

Following an expedition to Peru, Barcroft (1925) came to the conclusion that there was a tendency for the blood O_2 affinity to increase after high-altitude exposure. Hall et al. (1936) studied the Bolivian goose, or huallata, the South

American ostrich, and the Andean camelids llama and vicuña, and concluded "it is hardly a coincidence that two native birds as well as two of the native mammals have blood with significantly greater affinity for oxygen than members of their class at sea level." In the very same year, however, Keys et al. (1936) noted a decrease in the affinity of hemoglobin for oxygen at high altitude, while Aste-Salazar and Hurtado in 1944 advocated the advantage of a right-shifted OEC. Since that time, many studies have addressed the problem. Directly quoting from L.R.G. Snyder et al. (1982, pp 90–91), these studies have yielded "a general impression reiterated by numerous reviewers (for instance, Bullard 1972; Lenfant 1973; Bartels and Baumann 1977; Bunn 1980) that the responses to hypoxic stress are fundamentally different in animals native to low versus high altitudes. According to that simple dichotomy, in mammals native to low altitude, acclimation to high altitude or other hypoxic stress generally involves an increase in P_{50} ... In contrast, mammals and birds considered native to high altitude purportedly are adapted by virtue of lower than normal P_{50}'s. Furthermore, it appears that in those animals the P_{50}'s do not change significantly upon acclimation to different altitudes."

a) Acute and short-term hypoxia: A small increase in P_{50} measured in "standard" conditions is the most consistent finding in humans (Lenfant et al. 1968, 1971), in a wide variety of mammals, including dogs, cats, rabbits, guinea pigs, and rats (references in Snyder 1982), as well as in birds (Black et al. 1978) exposed to high altitude. Figure 4.4A indicates that this increase occurs rapidly after the onset of hypoxia, and declines within a few days upon cessation of hypoxia. Its magnitude would be related to the altitude and thus to the severity of hypoxia

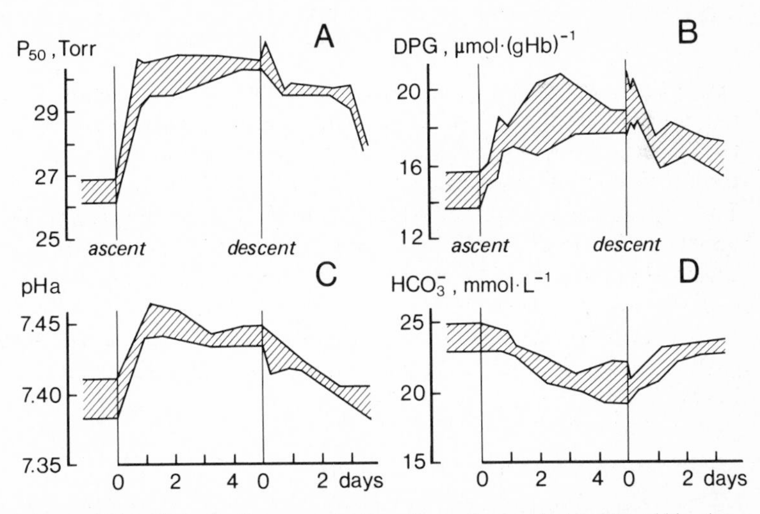

Fig. 4.4. Changes in hemoglobin affinity for O_2, concentration of organic phosphate, pH, and bicarbonate in blood of sea-level natives after ascent to and descent from the altitude of 4.51 km. (Derived from Lenfant et al. 1971). *Abscissae* time. *Ordinates: A* partial pressure of O_2 corresponding to 50% saturation of hemoglobin with O_2 (P_{50}), in vitro at 37 °C and plasma pH = 7.4; *B* concentration of 2,3-diphosphoglycerate (DPG); *C* pH in the arterial blood (pHa); *D* bicarbonate concentration in the arterial blood (HCO_3^-). Ranges of values are shown for six normal subjects by *hatched areas*. In all subjects, arterial P_{CO_2} decreased from 40 to 32–28 Torr at high altitude due to the hypoxic hyperventilation

(Torrance et al. 1970/71). Not all investigators, however, found a significant increase of P_{50} (Weiskopf and Severinghaus 1972).

By contrast, a decrease in P_{50} is typical in fish acclimated for a few weeks to hypoxic waters (Wood and Johansen 1972; Soivio et al. 1980; Tetens and Lykkeboe 1981, Table 4.2). Similarly, it seems that high altitude amphibians have a higher blood O_2 affinity than their relatives from lowland areas (Hutchinson et al. 1976). Nothing appears to be known in reptiles in this respect (Pough 1980).

b) Long-term hypoxia: For human beings, no clear difference exists between the P_{50} values in high-altitude natives and those in sea-level dwellers at identical

Table 4.2. Blood O_2 affinity, estimated by the O_2 partial pressure at which 50% of the hemoglobin is oxygenated (P_{50}), and organic phosphates in red cells in humans and animals at various altitudes or levels of oxygenation

Species	Condition	P_{50} Torr	Phosphates		Ref.
			Nature	Concentration	
	Altitude, m			mol · (mol Hb)$^{-1}$	
Humans					
Peruvians	0 native	26.6	2,3 DPG	0,61	1
–	4500 acute	31.1	–	0.94	
–	4500 accl.	27.3	–	0.57	
–	4500 native	30.7	–	0.80	
–	0 native	26.7			2
–	4540 native	31.2	–	1.15	3
Italians	120 native	26.5	–	0.97	4
–	3500 accl.	28.2	–	1.22	
Sherpas	3850 native	27.3	–	1.22	
Camelids					
Llama	0 accl.	22.8			5
–	3420 native	23.7			
Camel	0 native	23.3			6
				mmol · (L red cells)$^{-1}$	
Birds					
Canada goose	0 native	42.0	sum	9.1	7
Bar headed goose	0 –	29.7	–	7.2	
	P_{O_2}, Torr			mol · (mol Hb)$^{-1}$	
Ectotherms					
Trout 20 °C	150	24.1	ATP	1.31	8
– –	80 accl.	21.7	–	1.00	
– –	30 –	16.8	–	0.54	
Carp 20 °C	120–150	8.6	–	1.31	9
– –	30 accl.	3.8	–	1.04	

The values of P_{50} were determined in vitro (at 37 °C for the listed mammals and birds and at 20 °C for the ectotherms), and standardized to the following pH values: 7.2 for birds, 7.4 for humans and camelids, 7.8 for ectotherms

References: 1. Lenfant et al. (1968); 2. Winslow et al. (1978); 3. Winslow et al. (1981); 4. Samaja et al. (1979); 5. Banchero et al. (1971); 6. Bartels et al. (1963); 7. Petschow et al. (1977); 8. Tetens and Lykkeboe (1981); 9. Weber and Lykkeboe (1978)

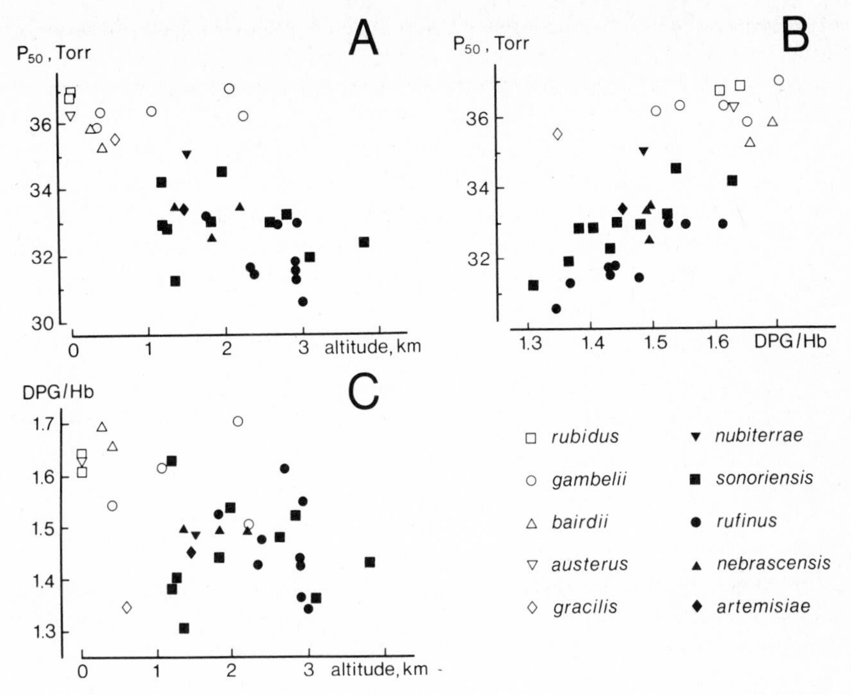

Fig. 4.5. Blood O_2 affinity in high- and low-altitude populations of the deer mouse, *Peromyscus maniculatus*. (Derived from L.R.G. Snyder et al. 1982; Snyder 1982). Ten subspecies of deer mice (*listed right*), which vary in their native altitudes, were studied after prolonged acclimation to low altitude (0.34 km for 3–30 months). *A*: P_{50} (determined at 37 °C, $P_{CO_2}=40$ Torr and pH = 7.4) versus native altitude of the population; *B*: P_{50} versus DPG/Hb molar ratio; *C*: DPG/Hb molar ratio versus native altitude of the population

elevations (Table 4.2). Studies on Andean natives, (Winslow et al. 1981) stress two points: (1) at the altitude of 4.54 km, blood O_2 affinity in vivo is not lower than that of sea-level controls, (2) considerable variation exists in P_{50}, both at sea level and high altitude. Horvath and Jackson (1982), in reviewing the literature, also found a great variability in the P_{50} measurements. In Himalayan natives, low values of P_{50} have been reported in Sherpas (Morpurgo et al. 1976), but the observation was not confirmed in a later study (Samaja et al. 1979, Table 4.2). It is not clear whether this variability in results reflects ecological differences (Clench et al. 1981), different kinds of altitude adaptation, simply individual variations, or incompatibility of the methods employed.

In mammals and birds (Table 4.2 and Fig. 4.5 A) the values of P_{50} appear to be lower in animals native to high altitudes than in their relatives from lowland habitats. A notable exception is the llama, which has a P_{50} insignificantly different from that of its near relative the camel (Table 4.2). It could hence be that the high blood O_2 affinity of the llama (Chiodi 1970/71) is simply a genetic trait of the camelidae, unrelated to high-altitude adaptation (Bullard 1972).

To evaluate correctly a possible adaptive significance of blood O_2 affinity, L.R.G. Snyder et al. (1982, p 92) have advocated the need of "studies in which

(1) comparisons are made among animal taxa that are as closely related as possible, (2) the evolutionary histories of the animals vis-a-vis high altitude can be assessed with some confidence, and (3) genetic factors potentially contributing to high altitude can be identified." These authors have initiated studies on population genetics and physiology of the deer mouse, *Peromyscus maniculatus,* a species with numerous closely related subspecies which vary in their altitudinal ranges. They found (Fig. 4.5 A) a strong negative correlation between P_{50} and native altitude. Tests on progeny reared at low altitude suggested that the differences in P_{50} were primarily genetic. On this basis, we may conclude, at least for the deer mouse, that an increased blood O_2 affinity might be a truly adaptive character related to high altitude hypoxia. Further studies of similar nature must substantiate or refute this trend.

c) Effectiveness of OEC shifts: Figure 4.6 compares the blood O_2 transport in llamas, with high O_2 affinity, to that in sheep, with low O_2 affinity, at different altitudes in a hypobaric chamber. The data (Banchero and Grover 1972) clearly suggest a beneficial effect of low blood O_2 affinity under mild hypoxia (1.6–2.8 km); conversely, a high blood O_2 affinity is advantageous under more severe hypoxia (4.5–6.4 km). As pointed out by the authors, llamas tolerate the highest elevation studied with no signs of discomfort, while sheep showed marked cyanosis, hyperventilation, and agitation.

Several other studies support the view that an increased blood O_2 affinity may be an important factor in adaptation to altitude hypoxia. These studies, however, have solely demonstrated an increased *survival* during acute exposure at extreme simulated altitude. Thus, Eaton et al. (1974) found that Sprague-Dawley rats with a cyanate-induced[4] left shift of the OEC survived, and had an uneventful recovery from 90 min exposure to an equivalent altitude of 9.18 km in a hypobaric chamber, whereas eight of ten control animals died. Similar results were obtained by Penney and Thomas (1975), and by Turek et al. (1978 b, c). The latter study, performed on anesthetized animals, confirms the previous theoretical analysis by Turek et al. (1973 a) by showing that severe hypoxia ($Pa_{O_2} < 31$ Torr) resulted in differences between venous P_{O_2} and arterio-venous O_2 concentration greater in NaOCN-treated rats of Wistar strain (high blood O_2 affinity) than in controls (low blood O_2 affinity). At moderate hypoxia (with Pa_{O_2} close to 60 Torr), a reversed situation was observed. A noteworthy finding was that, at an intermediate level of oxygenation ($Pa_{O_2} = 50$ Torr), the left-shifted OEC of NaOCN-treated rats neither impaired nor improved muscle oxygenation, judging from the skeletal muscle surface O_2 pressure fields (Nylander et al. 1983).

For human beings, Hebbel et al. (1978) also support the idea that not a right-shifted OEC, as most often reported, but a left-shifted OEC is of adaptive significance at moderate altitudes. These authors studied two healthy adolescents from a family strain having a rare high O_2-affinity mutant hemoglobin (Hb Andrew-Minneapolis; whole blood $P_{50} = 17$ Torr). In comparison with two normal sib-

4 For 2 weeks prior to study, the animals were given drinking water containing 0.5% sodium cyanate (NaOCN), a substance known to irreversibly carbamylate hemoglobin amino groups, thereby increasing Hb-O_2 affinity. After this treatment, the "standard" P_{50} value was 21 Torr, against 37 Torr in control rats, which were given drinking water with 0.49% NaCl to equalize sodium intake in the two groups

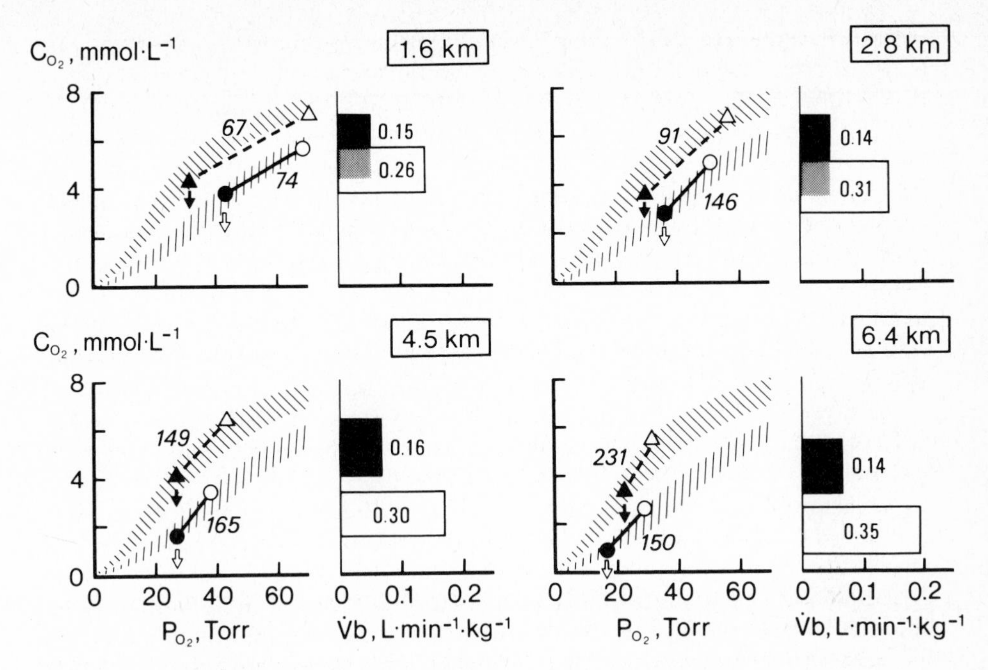

Fig. 4.6. O_2 transport by blood in llamas (*triangles* and *closed rectangles;* high blood O_2 affinity; P_{50} = 23 Torr) and sheep (*circles* and *open rectangles;* low blood O_2 affinity; P_{50} = 40 Torr) at different altitudes simulated in a hypobaric chamber. (Derived from Banchero and Grover 1972). Four panels are shown, as identified by the altitude indicated in the *top right hand corner* of each panel. Every panel is organized as Fig. 4.1 and comprises two representations. *In each representation of* C_{O_2} *vs.* P_{O_2}, the *open symbols* refer to arterial blood, and the *closed ones* to mixed venous blood; the *slopes of the lines* joining related arterial and mixed venous points, and *associated figures in italics,* indicate the blood O_2 capacitance coefficient [$\mu mol \cdot L^{-1} \cdot Torr^{-1}$; βb_{O_2} in Eqs. (4.2) and (4.3)]; the *sigmoid hatched areas* show schematically the range of the blood O_2 equilibrium curves at 38°–39 °C for the various pH encountered during the experiments. *In each representation of* C_{O_2} *vs.* $\dot{V}b$, the cardiac blood flow ($\dot{V}b$) and the oxygen consumption ($\dot{M}_{O_2}$; *rectangles* and *related figures*) are expressed per unit body mass; the differences in $\dot{M}_{O_2}$ and $\dot{V}b$ between the two animal species relate, at least in part, to the difference in body mass (llamas = 110–184 kg; sheep = 29–38 kg). When going from the altitude of 1.6 km to 6.4 km, note the following points: (1) arterial P_{O_2} was consistently higher in llamas than in sheep; (2) blood O_2 capacitance coefficient increased more in llamas than in sheep; (3) P_{O_2} in mixed venous blood was lower in llamas (*closed arrows*) than in sheep (*open arrows*) at moderate altitudes (1.6 and 2.8 km = mild hypoxia), but higher at the greatest elevation (6.4 km = severe hypoxia); (4) accordingly, the magnitude of the decline in mixed venous P_{O_2} was small in llamas (8 Torr), but large in sheep (26 Torr)

lings (P_{50} = 27 Torr), the high affinity subjects manifested similar hyperventilation, alkalotic changes in pH, and increases of both hematocrit and hemoglobin concentration consequent to the translocation from low (0.25 km) to high altitude (3.1 km). However, when exercising at high altitude, normal subjects showed the expected decrement in the maximal O_2 consumption (about 20%; Fig. 2.5), whereas high affinity subjects showed no decrement. Hebbel et al. (1978) concluded that subjects with high oxygen affinity hemoglobins may be preadapted to altitude. They wrote "Because hominids evolved at low altitudes, exposure to hypoxia has most frequently been due to a decrement in oxygen-carrying capacity (anemia) rather than a limitation in the availability of oxygen (hypoxic hypoxia). In the former circumstance, a right shift of the OEC is unquestionably advantageous; it facilitates oxygen unloading without compromising uptake from the oxygen-rich environment. In contrast, water-breathing animals are most often made hypoxic through environmental oxygen deprivation (e.g., due to temperature-dependent changes in the solubility of oxygen in water). In this situation, a left shift of the OEC is appropriate because it facilitates oxygen loading in the gill... In terms of the logistics of oxygen loading and unloading, the human at altitude is more like the fish in an oxygen-poor environment than like the sea-level human with anemia. Despite this, the human response to both types of hypoxia is the same: a right shift of the OEC. Thus, humans may respond inappropriately to altitude exposure. Indeed, it is unreasonable to assume that an organism which lacks evolutionary exposure to an oxygen-poor environment should necessarily respond appropriately to altitude-induced hypoxia."

The importance of a right shift of the OEC on tissue oxygenation has been questioned for some time. On the basis of a critical analysis of the various factors involved in the oxygen transport at high altitude, Torrance et al. (1970/71) concluded that the right-shifted OEC (increased P_{50}) observed in both newcomers and permanent residents at 3.8–4.5 km altitude had insignificant effect on the mixed venous P_{O_2}. Dempsey et al. (1975) also found that right shifts of the OEC have either very small or even negative net influences on O_2 delivery to skeletal muscles in healthy man during prolonged muscular work following short-term adaptation at an altitude of 3.1 km.

4.2.2.4 Underlying Mechanisms

Despite considerable progress in recent years, the manner in which blood O_2 affinity in vivo is controlled under hypoxic condition is not fully understood. There is evidence for an intricate interplay between (1) biochemical mechanisms within the red blood cells, (2) physiological influences arising from outside the cells, (3) genetically determined intrinsic properties of hemoglobin and concentrations of cofactors. The factors included in items 1 and 2 will be examined together under the following subheadings (a) effects of low P_{O_2} and (b) effects of akalosis. Some other interacting factors and genetic influences will be considered under subheadings c and d, respectively.

a) Effects of low P_{O_2}: It bears stressing that oxidative phosphorylation is the primary energy source in nucleated red blood cells of all nonmammalian vertebrates, whereas anaerobic glycolysis is the sole energy source in mammalian red cells. The functional consequence is that hypoxia may induce a decrease of adenosine triposphate (ATP) in nucleated red cells (Fig. 4.7, top left hand corner), whereas it does not in mammalian red cells (Fig. 4.8). Since organic phosphates (hence ATP) are allosteric effectors that bind preferentially to the deoxygenated conformation of hemoglobin, thereby decreasing Hb-O_2 affinity, it follows that a decrease of ATP in nucleated red cells of nonmammalian vertebrates will increase blood O_2 affinity. Such a mechanism is present in fish exposed to hypoxia

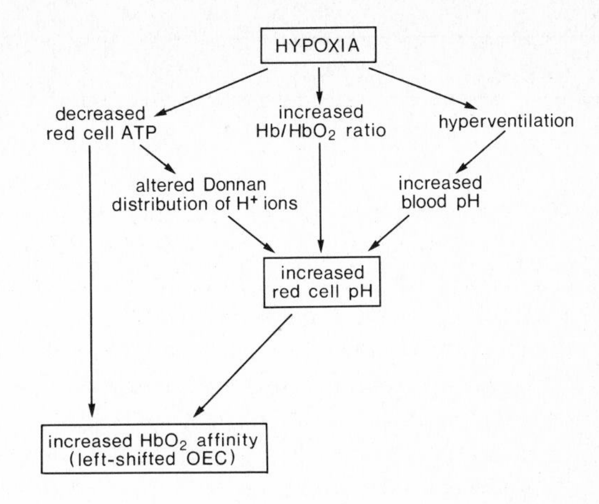

Fig. 4.7. Mechanisms controlling ATP concentration and Hb-O_2 affinity in nucleated red blood cells during hypoxia. (After Wood and Lenfant 1979)

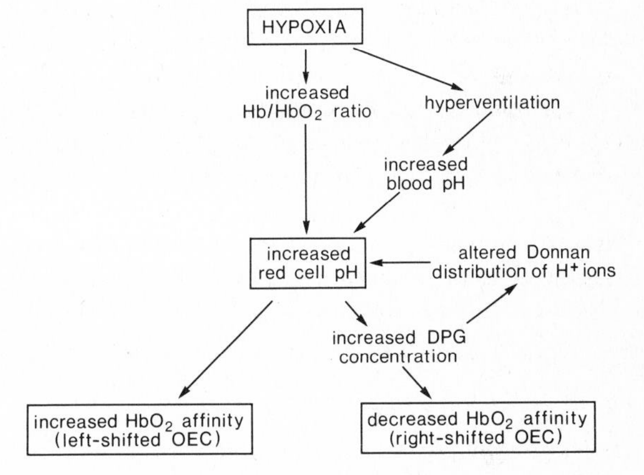

Fig. 4.8. Mechanism controlling concentration of 2.3-diphosphoglycerate (DPG) and Hb-O_2 affinity in mammalian red cells. (After Duhm and Gerlach 1971)

(Wood and Johansen 1972, Table 4.2). ATP concentrations in avian erythrocytes may be similarly controlled (Ingermann et al. 1983). As Fig. 4.7 indicates, the change in ATP concentration is thought to affect blood O_2 affinity not only through a direct (allosteric) effect, but also through the Bohr effect because of altered Donnan equilibrium of ions (including H^+) across the red cell membrane, and the resulting increased pH.

In all vertebrates, a primary effect of hypoxia is the reduction of blood O_2 saturation. Since deoxygenated hemoglobin (Hb) is more alkaline than HbO_2, the red cell pH increases, thereby favoring the increase of blood O_2 affinity (Figs. 4.7 and 4.8).

b) Effects of alkalosis: Another primary effect of hypoxia is to provoke hyperventilation, hypocapnia, and alkalosis. This, due to the Bohr effect, tends to increase blood O_2 affinity in all vertebrates (Figs. 4.7 and 4.8). In mammals, however, it is currently thought that the increased red cell pH yields the opposite effect, i.e., a decreased blood O_2 affinity, through the allosteric action of 2,3-diphosphoglycerate (DPG[5] in Fig. 4.8), the concentration of which has been reported to increase within a few hours under altitude hypoxia (Fig. 4.4B; Lenfant et al. 1971; Gerlach and Duhm 1972; Snyder 1982). Depending on the respective contributions of the opposite Bohr and DPG effects, the blood O_2 affinity may be affected to varying extents, or left unchanged.

c) Interacting influences may be responsible for seemingly contradictory observations. For instance, it is known that the O_2 affinity of hemoglobin is concentration-dependent (see reviews already quoted). In vivo experiments, discussed by Shappell and Lenfant (1975) indicate that decreasing the mean corpuscular hemoglobin concentration (MCHb in Fig. 4.2) yields higher blood O_2 affinity at given DPG/Hb molar ratio, and reduced sensitivity of hemoglobin to change in the concentration of its allosteric effector. The extent to which this effect opposes that of increasing levels of DPG under hypoxia is not precisely known, although a strong negative correlation between MCHb and DPG concentration was observed in humans at the altitude of 3.1 km (Brewer et al. 1972). Differences in hormonal influences (Brewer et al. 1972) and ecological factors (Clench et al. 1981) appear to be important; when not controlled, they may explain some discrepancies among P_{50} determinations.

d) Genetic influences determine not only the amino acid sequences of hemoglobins, which vary among animals, but also the nature and the intraerythrocytic concentrations of organic phosphates, and the interactions of hemoglobins with these allosteric effectors. Thus the high blood O_2 affinity of some animals normally living at high altitude, such as the llama guanaco or the bar headed goose, can be explained by two factors: reduced red cell concentration of organic phosphates[6] and hemoglobins having lower reactivity toward organic phosphates (Petschow et al. 1977). Similarly, the blood from embryos of highland geese has higher O_2 affinity than that from embryos of lowland relatives; the difference appearing to reside in the intrinsic properties of hemoglobin itself or in the interaction of hemoglobin with the organic phosphates that modulate O_2 affinity (G.K. Snyder et al. 1982b). The deer mice native to high altitude studied by L.R.G. Snyder (1982) showed lower DPG/Hb ratios than did low-altitude natives (Fig. 4.5C), a difference that appears to be determined primarily by genetic factors. As recently reviewed by Bunn (1980), many of these phenomena can be satisfactorily explained at the molecular level, but their adaptational significance is less clear.

---

5 DPG is formed in a side branch of the glycolytic sequence; it originates from 1,3-phosphoglycerate (1,3-PG in Fig. 6.2) through the action of DPG-mutase, and returns to 3-phosphoglycerate (3-PG) through the action of DPG-phosphatase. The increased red cell pH is thought to stimulate the enzymes phosphofructokinase (which increases glycolytic rate) and DPG-mutase, and to inhibit DPG-phosphatase. Prevention of the alkalosis caused by the ventilatory response to hypoxia with acetazolamide (Lenfant et al. 1971) or CO_2 inhalation (Gerlach and Duhm 1972) leaves the normoxic DPG concentration unchanged. The DPG synthesis would be controlled by two negative feedback mechanisms (1) the product inhibition of DPG-mutase, (2) the return of the red cell pH toward lower values, due to the accumulation of DPG which, as a non penetrating anion, changes the Donnan equilibrium across the red cell membrane (Duhm and Gerlach 1971)

6 DPG in guanaco blood, and IPP (adenosine inositol pentaphosphate, the primary effector in avian erythrocyte) in the bar headed goose. In contrast, the concentration of these organic phosphates rises in the animals experimentally subjected to altitude hypoxia

4.3 Blood Flow

Apart from the quantity and the quality of hemoglobin examined in the foregoing section, it makes immediate sense that changes in blood flow [$\dot{V}b$ in Eqs. (4.1), (4.2), and (4.3)] may be important for the O_2 transport to tissues (arrow labeled "flow reserve" in Fig. 4.1). This consideration is valid not only for the whole body and cardiac output, but also for any organ and its own blood flow rate. Redistribution of blood away from some areas (e.g., the skin) to others (e.g., active muscles) can make the body able to function more efficiently in the face of hypoxia. Contrarily, if its blood flow rate does not increase when its activity and oxygen demand increase, any organ or tissue is compelled to extract more O_2 from the inflowing blood, at the expense of a decrease in both O_2 tension and concentration in the outflowing blood (arrow labeled "blood reserve" in Fig. 4.1). The consequence is a lowered pressure head for O_2 diffusion to cells; additional tissue capillaries must open or be developed for the reduction of the diffusion path and for a new equilibrium to be achieved (arrow labeled "capillary reserve" in Fig. 4.1). Although these adjustments are interrelated, what follows concentrates on the cardiac output and its distribution; possible changes in capillarization will be examined in Chap. 5.

4.3.1 Cardiac Output

Since the beginning of this century, the effects of high altitude on the cardiac output ($\dot{V}b$, the minute volume output from each ventricle) and its components, frequency of systole (f_H) and stroke volume (Vs, the volume of blood ejected during systole), have been investigated by many workers, essentially in man (see Grollman 1930; Stickney and van Liere 1953; Balke 1964; Pugh 1964). Not all observations are in agreement, especially in the early phase of exposure to hypoxia. Salient features, however, can be summarized as follows.

4.3.1.1 Resting Conditions

The most readily apparent effect of altitude hypoxia is tachycardia. Heart (systolic) frequency (f_H; Fig. 4.9) increases in proportion with the severity of hypoxia (a phenomenon quickly reversed after inhalation of oxygen), but returns to almost sea-level values as acclimatization proceeds. A decreased cardiac stroke volume (Vs, usually estimated from the cardiac output and heart frequency) is also a common feature under altitude hypoxia, reported in humans and other mammals, as well as in birds (e.g., Sime et al. 1974; Levasseur et al. 1976; Hoon et al. 1977; Hannon and Vogel 1977; Besch and Kadono 1978; Bouverot et al. 1976, 1979). Underlying mechanisms are not fully understood. Sympathetic stimulation and release of catecholamines in plasma (Cunningham et al. 1965) are believed to be responsible for the observed tachycardia. Local depressant effect of hypoxia and hypocapnic alkalosis are possibly involved in the decrease of the stroke volume (Grover et al. 1976a).

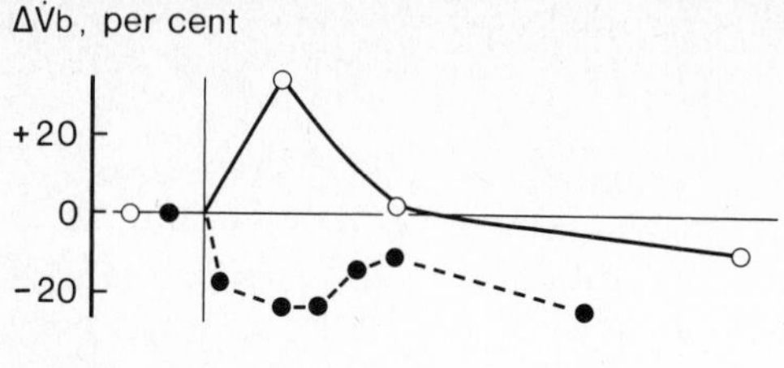

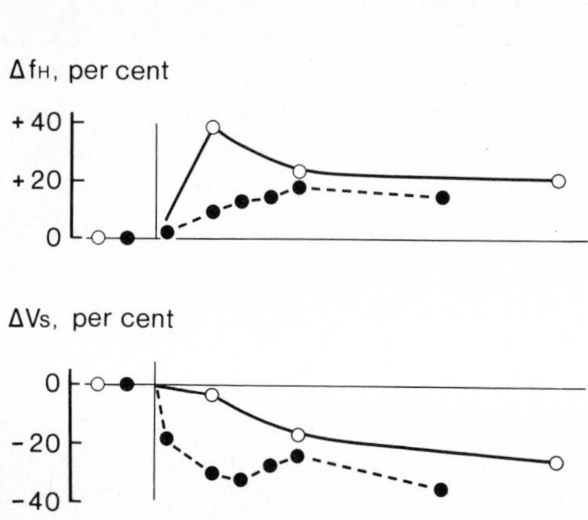

Fig. 4.9. Relative changes in resting cardiac output ($\dot{V}$b), heart frequency (fH) and cardiac stroke volume (Vs) of healthy sea-level residents in the course of exposure to the altitude of 4.3 km (*open symbols* Hannon and Vogel 1977) or 3.7 km (*closed symbols* Hoon et al. 1977)

Since $\dot{V}b = Vs \cdot fH$, the cardiac output may theoretically incrase, or decrease, or still remain unaltered under hypoxia, depending on the relative importance of the opposite changes in its components. For instance, in calves exposed at the altitude of 3.4 km, unchanged fH but decrased Vs were found to result in a significant reduction of $\dot{V}$b (Ruiz et al. 1973). In the other mammals studied so far, $\dot{V}$b most often increased in the early period of hypoxia, due to tachycardia, but returned to preexposure values within a few days, or even fell to levels that are below those observed before exposure to hypoxia, due to the secondary reduction in fH and Vs (Hultgren and Grover 1968; Grover et al. 1976 b). Yet the reduction in Vs might be such that $\dot{V}$b started to fall immediately on arrival at high altitude (Hoon et al. 1977, Fig. 4.9).

In resting healthy high-altitude natives, the cardiac output has been reported as normal rather than diminished. Thus, Torrance et al. (1970/71) found no difference in the ratio $\dot{V}b/\dot{M}_{O_2}$ (i.e., the blood convection requirement; line 17 in Table 4.3) when comparing sea-level and high-altitude natives studied at their native elevation. For the two groups of individuals, Table 4.3 shows the values of all the variables involved in Eqs. (4.1), (4.2), and (4.3). Taking into consideration that the high altitude group had lower body mass (line 3) and lower O_2 consumption (line 7), note that the $\dot{M}_{O_2}$-specific conductance between arterial and mixed venous blood ($Ga,\bar{v}_{O_2}/\dot{M}_{O_2}$ in line 18) was 4.5 greater in the high altitude natives. This resulted not from a change in the ratio $\dot{V}b/\dot{M}_{O_2}$, but from the much higher blood O_2 capacitance (line 15; βb_{O_2} was 4.24 times greater). This in turn related to two factors: (1) the arterio-venous C_{O_2} and P_{O_2} differences were located on a

Table 4.3. Blood O_2 transport in Peruvians born and raised either near sea level (s.l.) or at the altitude of 4.34 km (h.a.), and studied at their native altitude. (Torrance et al. 1970/71). Mean values from 6 subjects in each group (italic numbers = 1 SE)

		Sea level	High altitude	h.a/s.l.
1 Elevation	m	10	4340	
2 Barometric pressure	Torr	760	458	
3 Body mass	kg	70.1	58.6	0.84
4 Hematocrit	%	46.7	54.4	
5 Hemoglobin conc.	$g \cdot L^{-1}$	164	193	1.18
6 P_{50} (pH = 7.4)	Torr	26.7	30.5	
7 $\dot{M}_{O_2}$	$mmol \cdot min^{-1}$	12.63	10.85	
		0.67	*0.63*	
8 Pa_{O_2}	Torr	93.9	45.2	0.48
9 $P\bar{v}_{O_2}$	Torr	39.4	32.9	0.84
10 $Pa_{O_2}-P\bar{v}_{O_2}$	Torr	54.5	12.3	
11 Ca_{O_2}	$mmol \cdot L^{-1}$	9.31	8.41	0.90
12 $C\bar{v}_{O_2}$	$mmol \cdot L^{-1}$	7.05	6.27	0.89
13 $Ca_{O_2}-C\bar{v}_{O_2}$	$mmol \cdot L^{-1}$	2.26	2.14	
14 Eb_{O_2}		0.24	0.25	1.04
15 βb_{O_2}	$mmol\ (L \cdot Torr)^{-1}$	0.041	0.174	4.24
16 $\dot{V}b$	$L \cdot min^{-1}$	5.58	5.08	
		0.36	*0.33*	
17 $\dot{V}b/\dot{M}_{O_2}$	$L \cdot mmol^{-1}$	0.442	0.468	1.06
		0.011	*0.025*	
18 $Ga,\bar{v}_{O_2}/\dot{M}_{O_2}$	$Torr^{-1}$	0.018	0.081	4.50

part of the blood O_2 equilibrium curve which was much steeper at high altitude than at sea level, (2) the concentration of hemoglobin (line 5 in Table 4.3) was 18% higher in the high-altitude natives than in sea-level subjects; consequently, the blood O_2 equilibrium curve was steeper in the former than in the latter. The functional consequence was that the venous P_{O_2} (line 9), hence the pressure head for O_2 diffusion to cells, was well preserved in the high-altitude subjects studied in resting conditions by Torrance et al. (1970/71). Due to the high blood O_2 capacitance in these subjects, there was no need for the costly process of increasing cardiac output.

For animals native to high altitude, an identical conclusion can be drawn from the data obtained by Banchero and Grover (1972) in the llama studied at various elevations (Fig. 4.6). Similarly, the blood convection requirement ($\dot{V}b/\dot{M}_{O_2}$) remains unaffected in the bar headed geese when these birds are exposed to altitudes up to 9.15 km (Black and Tenney 1980).

In water-breathers also, the cardiac output seems to be well maintained under hypoxia. However, in great contrast to mammals and birds, hypoxic fish exhibit a marked bradycardia, which is offset by an increase in stroke volume (Holeton and Randall 1967; Wood and Shelton 1980).

The mechanisms concerned with cardiac regulation under hypoxia are complex, and fully understood neither in mammals (Korner 1971; Krasney and Koehler 1980) nor in fish (Wood and Shelton 1980; Randall 1982). In mammals, afferent signals from the lungs may reflexly contribute in part to the cardioaccelerator response. Thus, in dogs, Daly and Scott (1963) showed that the tachycardia associated with stimulation of the hypoxic carotid body could be reversed to bradycardia by applying steady artificial ventilation, thereby preventing the hypoxic hyperventilation and concomitant hypocapnia. Prevention of hypocapnia by adding some CO_2 to the inspired gas decreased, but did not reverse tachycardia. Hence, the primary cardio-inhibitory response via the carotid body appears to be antagonized in aerial vertebrates, at least in dogs, by the ventilatory changes induced by hypoxia.

4.3.1.2 Exercise

In running dogs a higher transfer of oxygen from the lungs to the tissues ($\dot{M}_{O_2}$) is made possible by increases of the blood flow rate ($\dot{V}b$) and of the arterio-venous O_2 concentration difference ($Ca_{O_2} - C\bar{v}_{O_2}$), according to the Fick Eq. (4.1) (Fig. 4.10). These variables increase with increasing levels of exercise, and reach maxi-

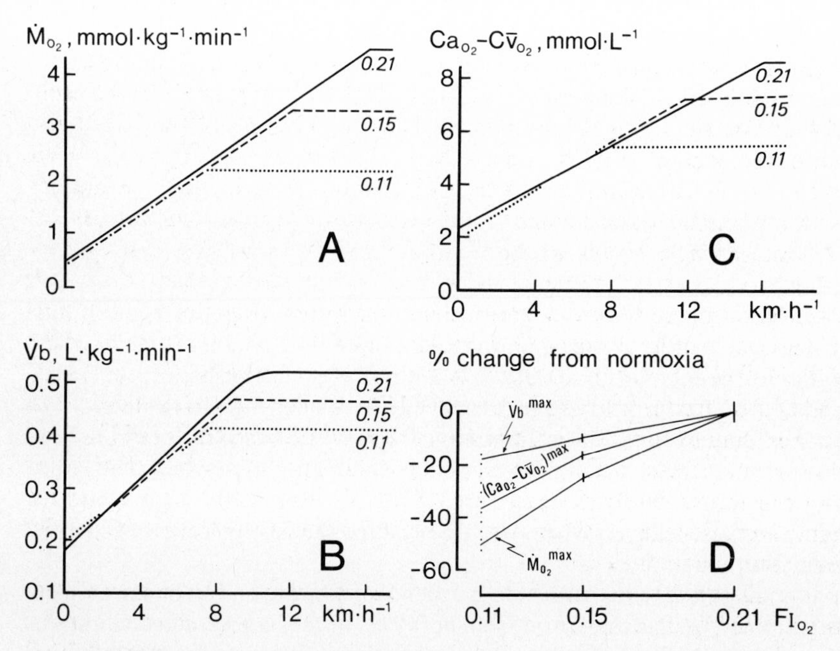

Fig. 4.10. Effects of hypoxia upon O_2 consumption ($\dot{M}_{O_2}$), cardiac output ($\dot{V}b$), difference in concentration of O_2 in arterial and mixed venous blood ($Ca_{O_2} - C\bar{v}_{O_2}$) in dogs at rest and running on a treadmill with an incline of 10% at speeds varying from 4 to 16 $km \cdot h^{-1}$. (After Piiper et al. 1966). In *A*, *B*, and *C*, the dogs were at sea level and breathed either room air (*solid lines* $F_{O_2} = 0.21$) or gas mixtures containing 15% or 11% O_2 in nitrogen simulating acute exposure to the altitudes of 2.7 and 5 km, respectively (*dashed lines* $F_{O_2} = 0.15$; *dotted lines* $F_{O_2} = 0.11$). $\dot{M}_{O_2}$ was determined by means of an open-circuit system connected to a respiratory mask, and $\dot{V}b$ by a thermodilution method; $Ca_{O_2} - C\bar{v}_{O_2}$ differences were calculated as the ratio $\dot{M}_{O_2}/\dot{V}b$. Note that these variables increased at first, and then plateaued with increasing levels of exercise. Their maximal values were reached at lower speeds when the level of oxygenation decreased. Percent changes from normoxia in these maximal values are shown in panel *D*; note that maximal $\dot{V}b$ is not reduced enough to explain the decrease in $\dot{M}_{O_2}^{max}$. The reduced maximal arterio-venous O_2 difference possibly related to a reduced O_2 transport to the active muscles and/or to a lesser O_2 extraction

mal values, which remain unchanged with further increases of the exercise levels. With hypoxia, these limits are lower, and they are reached at lower exercise levels. The reduction of the maximal O_2 consumption ($\dot{M}_{O_2}^{max}$) at high altitude has been already examined (Fig. 2.5). Related changes in the cardiac output are considered below.

a) In acute hypoxia, the crossing point of $\dot{V}b$ curves in Fig. 4.10 B indicates that, in dogs, the cardiac output is elevated at rest and medium speeds of running, but lowered at high speeds. In exercising men, the crossing point, if any, seems to be located at higher levels of exercise. At high altitude, indeed, the cardiac output and heart frequency are raised above sea-level values at any submaximal work load (Asmussen and Chiodi 1941; Dejours et al. 1963; Pugh et al. 1964; Stenberg et al. 1966; Saltin et al. 1968; Vogel et al. 1974 a; Cerretelli 1976); at maximal work load, they are almost identical to sea-level values (Asmussen and Nielsen 1955; Stenberg et al. 1966; Vogel et al. 1967).

b) Prolonged exposure to hypoxia reduces both the submaximal and maximal values of $\dot{V}b$, mainly through a reduction of the heart frequency. Thus, after 2–12 weeks at 4.3–5.8 km, submaximal values of cardiac output return to sea-level controls, while maximal values are 10–30% below controls, as are maximal heart frequency and maximal stroke volume (Pugh et al. 1964; Saltin et al. 1968; Vogel et al. 1974a; Cerretelli 1976; Horstman et al. 1980). Changes of similar magnitude, but in the opposite direction, have been observed in life-long residents at 4.35 km, when relocated for 2 weeks to sea level (Vogel et al. 1974b), hence indicating that maximal cardiac output is also reduced in high-altitude natives.

c) The underlying mechanisms and functional significance of the changes mentioned above are uncertain. The depressant effect of hypoxia has been advocated. Yet reduction of maximal cardiac output and heart frequency develop progressively in the course of hypoxic exposure, indicating that hypoxia per se is not the single factor involved. Furthermore, only modest improvements of maximal O_2 uptake and heart frequency follow the relief of hypoxia by a switch from air to pure O_2 breathing in subjects acclimatized at high altitude (Cerretelli 1976). Increased parasympathetic activity has been reported to play some role (Hartley et al. 1974). Cardio-inhibitory reflexes originating in the arterial chemoreceptors may be involved, since increased cardiac blood flow results from carotid body denervation (Bouverot et al. 1981).

Whatever the underlying mechanisms are, the reduction of the maximal cardiac output ($\dot{V}b^{max}$) develops progressively under altitude hypoxia, as does the polycythemic response (Sect. 4.2.1) and the related increase in the arterial blood O_2 concentration (Ca_{O_2}). These concomitant changes act in opposite directions on the maximal systemic O_2 transport ($\dot{V}b^{max} \cdot Ca_{O_2}$). As discussed by Cerretelli (1980), they may explain that the maximal O_2 uptake ($\dot{M}_{O_2}^{max}$ in Fig. 2.5) is similarly reduced in acute and in chronic hypoxia, in highlanders as in lowlanders.

Considering the possible hemodynamic effects of the increased hematocrit and blood viscosity associated with prolonged exposure to hypoxia, it has been argued that blood-letting should improve performance (Sect. 4.2.1.2). However, Horstman et al. (1980) disagree with this idea, since in volunteers acclimatized for 3 weeks at the altitude of 4.3 km, they found that removal of 450 ml of whole blood (fluid replaced) reversed the 10% decrease in $\dot{V}b^{max}$ observed before blood-letting, but the maximal O_2 uptake was reduced by about 8%. Moreover, the endurance time of running at submaximal work loads (0.85 $\dot{M}_{O_2}^{max}$) was reduced by 35%. Hence, in normal (unbled) subjects, the effects of increased Ca_{O_2} during high-altitude sojourn exceeded those of reduced $\dot{V}b^{max}$. It would, therefore, seem "unsound to submit climbers to blood-letting even at extreme altitude," as stressed by Sutton and colleagues (1983), who consider "unlikely that the normally acclimatized and hydrated climber will develop a hematocrit and blood viscosity that would impair O_2 delivery."

4.3.2 Redistribution of Blood Flow

Hypoxia may increase the sympathetic vasomotor outflow to visceral organs through central and reflex mechanisms (Korner 1971; Krasney and Koehler 1980). On the other hand, the basal tone and responsiveness of vascular smooth muscle to neural stimuli are influenced by a variety of local factors, such as O_2,

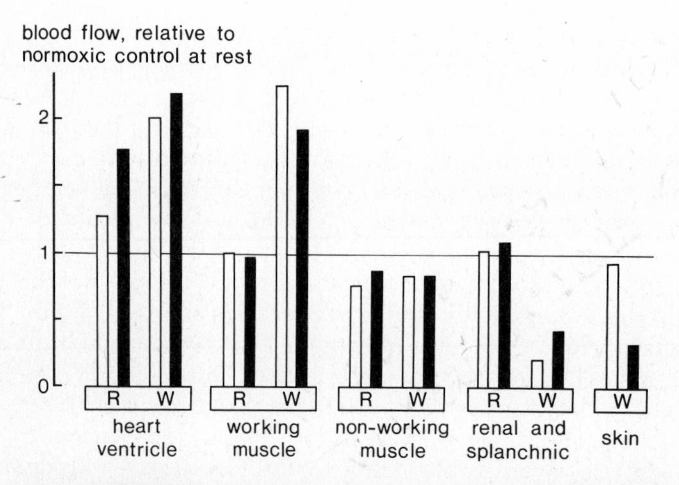

Fig. 4.11. Effects of acute hypoxia on regional blood flow at rest (*R*; ambient temperature = 22°–25 °C) and after swimming exercise (*W*; in water at 37 °C for 4 min) in Sprague-Dawley rats before (*open columns*) and after adaptation (*closed columns*) for 31–68 days (mean = 44 days) to the altitude of 4.4 km in a hypobaric chamber. (After Tucker and Horvath 1974). *Ordinate* regional blood flow determined as the fractional distribution of a radioactive indicator (cesium 137) per unit weight, and expressed as values relative to normoxic controls at rest. *Abscissa* various tissues and organs excised after 7–11 min of normobaric hypoxia (F_{O_2}=0.11; equivalent altitude = 5 km); *working muscle*: gastrocnemius, foreleg and scapular muscles; *non-working muscle*: muscles of the back and abdominal wall. The mean values of body mass, hematocrit and hemoglobin concentration in blood were respectively 373 g, 47.6% and 142 $g \cdot L^{-1}$ in sea-level animals, 340 g, 60.8%, and 194 $g \cdot L^{-1}$ in altitude-adapted ones. In the latter, blood lactate concentration after hypoxic exercise was lower than in sea-level controls (12.9 against 17.4 $mmol \cdot L^{-1}$). Note that (1) *under resting hypoxic condition (R)*, ventricular blood flow increased, while nonworking muscle blood flow decreased, (2) *during hypoxic swims (W)*, there was a marked increase in ventricular and working muscle blood flows, with concomitant marked reduction in renal and splanchnic blood flows, (3) *when comparing resting and working conditions in each group of rats*, a less pronounced redistribution of blood flows (skin excepted) occurred during exercise in the altitude-adapted animals

CO_2, H^+, K^+, Na^+, Ca^{2+}, catecholamines, temperature, osmolarity, etc. At the microcirculatory level, vascular smooth muscles are heterogenous in response to these factors, reflecting dynamic coupling between vascular function and metabolic rate (Duling 1978). Furthermore, vascular smooth muscle cells may be sensitive to changes in O_2 tension that are in the physiological range (Chang and Detar 1980). Conceivably, therefore, neurally mediated vasoconstriction can be overruled by local metabolic vasodilatation; regional variations in vascular resistances may divert blood flow from regions of low O_2 requirements to supply regions of higher demand, thereby allowing certain critical tissues to function more efficiently during hypoxia. In acute hypoxia, both the increase in cardiac blood flow and its regional redistribution may contribute to the delivery of O_2 to the various tissues (Nesarajah et al. 1983).

Figure 4.11 gives an example of such a redistribution of blood flow under altitude hypoxia. This figure summarizes some of the data obtained by Tucker and Horvath (1974) on rats exposed to acute hypoxia for the first time (open columns) or exposed to the same hypoxia after a 6-week period of acclimatization to the altitude of 5 km (filled columns), at rest (R) or during exercise (W). It is clear that the vascular beds of cutaneous, renal and splanchnic regions, and nonworking muscles (also adipose tissues; not shown) were the major source of the flow diverted to the heart and working muscles.

In the outline that follows, the experiments carried out at constant CO_2 tension on anesthetized, paralyzed, and artificially ventilated animals will not be considered, unless otherwise noted. Although providing valuable information about blood flow redistribution under *normocapnic hypoxia,* the data thus obtained do not apply to high-altitude *hypocapnic hypoxia;* due to the remarkable sensitivity of the vascular smooth muscle to CO_2, the balances between vasoactive and vasorelaxant influences reached in these two conditions are different. Unfortunately, data under hypocapnic hypoxia are scanty, and there are uncertainties about the methods for measuring blood flow currently in use. Furthermore, in most studies, only changes in blood flow to a single organ or tissue have been measured. Clearly, the magnitude and significance of redistribution of blood flow under altitude hypoxia cannot be outlined with accuracy at the present time.

4.3.2.1 Skin

As suggested in Fig. 4.11, the cutaneous circulation in a hot environment may act as a potential reserve for changing the distribution of flow during the process of acclimatization to high altitude. Using a plethysmographic method, Durand et al. (1969) measured blood flow, pressure and volume in the hand as a representative cutaneous vascular bed. They found that cutaneous blood flow decreased at high altitude (3.8–4.8 km), and was lower in higlanders than in new residents; in the two groups, the distensibility of forearm and hand capacitance was reduced by sea level standards. Extrapolating to the whole cutaneous circulation, Durand and Martineaud (1971) estimate that reduction in cutaneous flow induced by hypoxia may redistribute about 250 ml of blood when the skin temperature is above 33 °C. Hence, in humans, the effect of altitude hypoxia on skin perfusion should not play any significant role except under heat load, when the cutaneous flow is initially elevated.

In addition to arteriolar constriction, an increased venous tone may explain the above observations (Weil et al. 1969). Hypoxia itself (low P_{O_2}) seems to be responsible for the venoconstriction, and hypocapnia (low P_{CO_2}) for the arteriolar constriction (Cruz et al. 1976).

4.3.2.2 Renal and Splanchnic Areas

In resting, conscious rabbits and dogs (Vogel et al. 1969) and ewes (Nesarajah et al. 1983) the renal and splanchnic vascular beds were found to be the source of the flow fraction diverted to the heart and skeletal muscle. Yet these animals were studied acutely (20–30 min) under severe hypoxia ($F_{O_2}=0.08$, corresponding to 7.5 km altitude). Under less severe hypoxic challenge, Tucker and Horvath (1974, Fig. 4.11) observed no change in resting rats. In exercising animals, in contrast, the reduction of renal and hepatic blood flows was important, but not as great in the altitude-adapted rats as in controls. It is the opinion of these authors that the relatively higher perfusion of the liver and kidneys in the altitude-adapted animals, in association with the increased blood O_2-carrying capacity, provided adequate O_2 delivery to these organs, hence sustaining their lactate-metabolizing function.

In highlanders living permanently at La Paz (3.75 km), Capderou et al. (1977) found normal splanchnic blood flow (measured with indocyanine green). On the other hand, at Morococha (4.5 km), the renal plasma flow (RPF; derived from p-aminohippurate clearance) has been reported to be reduced by 40–50% in healthy natives compared with sea-level residents (Becker et al. 1957; Lozano and Monge 1965). These authors noted a 12–18% decrease in the glomerular filtration rate (GFR; measured by the clearance of inulin) and a 40–80% increase in the filtration fraction (i.e., the ratio GFR/RPF). An increased resistance of the efferent glomerular arteriole can explain the increased filtration fraction. This increased resistance, in turn, can be explained by an increase in the vascular tone induced by hypoxemia, by structural changes in the vessel, or by increased blood viscosity. Monge et al. (1969) have presented arguments supporting the view that the responsible factor is the increased viscosity: due to glomerular filtration, the hematocrit may increase dramatically in the efferent arteriole. Despite the probably very high hematocrit in the efferent arterioles, tubular function seems to be normal in high altitude natives (Monge et al. 1969).

4.3.2.3 Skeletal Muscle

The effect of hypoxia on muscular blood flow has been studied on skeletal muscle preparations (innervated or denervated) from anesthetized, artificially ventilated animals, with the arterial P_{CO_2} kept constant. In these normocapnic preparations, blood flow generally increases when decreasing arterial P_{O_2} below a critical value, e.g., 30 Torr in dogs (Cain and Chapler 1979), or 35–40 Torr in ducks (Grubb 1981). What changes are induced by hypocapnic hypoxia in these preparations are presently not known.

Measuring the rate at which intra-arterially injected xenon-133 was cleared from the tibialis anterior muscle, Bidart et al. (1975) found the muscular blood flow to be decreased (by 25% at rest, and 30% during exercise) in highlanders and sojourners at 3.8 km altitude, compared to lowlanders. Similarly, during hypoxic exercise in the rats studied by Tucker and Horvath (1974, Fig. 4.11), the blood flow to working muscles was increased less in the animals acclimatized at 5.0 km altitude than in controls. Thus, the perfusion to certain skeletal muscles would be reduced under prolonged hypoxia. Yet, taking into consideration the polycythemic response to hypoxia, the observed lower perfusion could no longer be significant if, instead of blood flow, the O_2 systemic transport (blood flow × arterial O_2 concentration) was calculated (Bidart et al. 1975).

4.3.2.4 Myocardium

Acute hypoxia appears to be accompanied by an increased ventricular blood flow, in sea-level controls as well as in animals previously acclimatized to high altitude. This has been demonstrated in rats using radioactive cesium-137 as an indicator (Tucker and Horvath 1974) and in calves using radionuclide-labeled microspheres (Manohar et al. 1982). Flow increases significantly under resting conditions (R in Fig. 4.11), concomitantly with increases in heart rate and right ventricular systolic pressure (Manohar et al. 1982); it increases further during hypoxic exercise (W in Fig. 4.11). These changes conform to the effects of acute hypoxia and exercise on coronary circulation recently reviewed by Feigl (1983). They may be more prominent in the animals previously acclimatized to high altitude (Fig. 4.11; filled columns compared with open ones; Manohar et al. 1982). Such an increment in ventricular myocardial blood flow in altitude-acclimatized animals has to be related to the cardiac hypertrophy, which is a common occurrence in rats (Ou and Smith 1983), in calves (Manohar et al. 1982) and in many other species, including mice, guinea-pigs, rabbits, cats, lambs, pigs, dogs, llamas, steers (references in Hultgren and Grover 1968), pigeons (McGrath 1971), and man (Heath and Williams 1981) exposed to prolonged hypoxia. The critical event is thought to be the right ventricular hypertrophy induced by the pressure overload related to hypoxic pulmonary vasoconstriction (Sect. 4.3.2.6).

Chronic hypoxia and its effects on myocardial circulation, apparently, remain to be studied in animals. In humans with right-heart catheterization, the coronary blood flow was found to be decreased (by about one third at 3.1–4.4 km) in high altitude natives (Moret 1971) and sojourners (Grover et al. 1976b), as compared with lowlanders. In the second study, in which the subjects were their own controls, the decrease in coronary blood flow was largely offset by a (28%) widening of the arterio-venous O_2 concentration difference in coronary blood; according to Eq. (4.1), therefore, the myocardial O_2 consumption was rather well maintained. Despite the decrease of the coronary sinus (venous) O_2 concentration, the corresponding P_{O_2} remained unaltered, around 18 Torr. Hence, a right shift of the blood O_2 equilibrium curve certainly occurred. The authors consider this lowered blood O_2 affinity as the initiating compensatory event that made the decrease in coronary flow the necessary consequence of the increased coronary O_2 extraction, not the primary cause of it.

4.3.2.5 Brain

Most studies on the effects of hypoxia on cerebral blood flow have related the flow to arterial O_2 tension (or saturation) while keeping the arterial CO_2 tension constant at its control value in normoxia. They have all demonstrated the vasodilator effect of hypoxia. Interestingly, Grubb et al. (1978) provided evidence suggesting that the ability of avian cerebral blood flow to increase at moderate levels of hypoxia may play a role in the exceptional tolerance of birds to hypoxia. However, since altitude hypoxia yields hypocapnia (Chap. 3.1.3), not normocapnia, and since the magnitude of the cerebral vasodilator response to hypoxia is largely dependent on the coexisting arterial P_{CO_2} (Shapiro et al. 1970), the studies referred to above will not be considered further.

The extent to which the vasodilator effect of hypoxia may be offset by the vasoconstrictor effect of hypocapnia under altitude hypoxia is not precisely defined. The upper part of Table 4.4 indicates that, in lowlanders acutely exposed to the altitude of 3.81 km, the cerebral blood flow (as determined with an N_2O method; Severinghaus et al 1966 b) increased by about 38%. After a few days of exposure, however, the cerebral blood flow returned to normal, due to hypoxic hyperventilation that induced hypocapnia and raised the arterial O_2 tension. Similarly, transiently increased retinal and, by inference, cerebral blood flow has been demonstrated by Frayser et al. (1974) following acute exposure to the altitude of 5.3 km; there was subsequently a significant decrease in flow after 9 days at high altitude.

In high-altitude natives who exhibit persistent hypocapnia (Table 4.4, lower part), the cerebral blood flow (studied with a krypton-85 wash-out method) was found to be reduced by about 25% (Durand et al. 1974). Sørensen et al. (1974) ob-

Table 4.4. Cerebral blood flow ($\dot{V}b^c$), arterial-venous O_2 concentration difference ($Ca_{O_2}-Cv_{O_2}$), brain oxygen uptake ($\dot{M}^c_{O_2}$), and O_2 and CO_2 partial pressures in arterial blood (Pa_{O_2}, Pa_{CO_2}) in man at sea level and at high altitude

	n	$\dot{V}b^c$ $\frac{L}{min \cdot kg}$	$Ca_{O_2}-Cv_{O_2}$ $\frac{mmol}{L}$	$\dot{M}^c_{O_2}$ $\frac{mmol}{min \cdot kg}$	Pa_{O_2} Torr	Pa_{CO_2} Torr	$\dot{V}b^c/\dot{M}^c_{O_2}$ relative to control
(1)							
Sea level	8	0.43	3.11	1.34	85.2	40.8	1.0
		0.02	*0.12*	*0.06*	*1.5*	*0.8*	
3.81 km, 6–12 h	6	0.58	2.37	1.32	43.5	35.0	1.38
		0.07	*0.18*	*0.06*	*1.4*	*0.9*	
3.81 km, 3–5 d	7	0.46	2.81	1.26	51.2	29.7	1.16
		0.03	*0.18*	*0.08*	*1.3*	*0.7*	
(2)							
Lowlanders	6	0.50	2.78	1.39	92.3	39.0	1.00
		0.01	*0.09*	*0.08*	*2.6*	*1.4*	
Highlanders at 3.8–4.8 km	16	0.40	3.78	1.50	59.7	28.9	0.75
		0.01	*0.17*	*0.06*	*1.2*	*0.5*	

1. Severinghaus et al. (1966); 2. Durand et al. (1974)
Mean values ± 1 SE (in italics), n = number of measurements

served a similar decrease in the altitude natives who had a high hematocrit, close to 50%. But in subjects with a hematocrit of 40%, representing slight anemia in highlanders, they found that the cerebral blood flow was the same as in normal man at sea level. Hence, increased hematocrit (erythropoietic reserve in Fig. 4.1), and therefore blood O_2 capacitance [βb_{O_2} in Eq. (4.2)], appear to determine the amount of O_2 supplied to the brain, in spite of the lower cerebral blood flow in the hypocapnic, otherwise healthy high-altitude natives.

4.3.2.6 Pulmonary and Branchial Circulation

In birds and mammals, pulmonary blood flow and the output of each ventricle represent virtually identical quantities, at least in steady state. Accordingly, the effects of hypoxia on the total pulmonary blood flow reflect those on cardiac output (Sect. 4.3.1). After a transient change in the early phase of exposure, the flow generally returns to or decreases below normal values in the course of acclimatization. However, pulmonary hypertension develops during hypoxia, whereas the systemic blood pressure is normal, or below normal in well-adapted subjects (see Heath and Williams 1981). The explanation is that sustained hypoxia exerts opposite effects on the systemic and pulmonary vascular smooth muscles, bringing about vasodilatation in the systemic circulation, but vasoconstriction in the pulmonary circulation. Numerous reviews have been concerned with hypoxic pulmonary vasoconstriction in the past decade (e.g., Fishman 1976; Lockhart and Saiag 1981; McMurtry et al. 1982); accordingly, only salient aspects will be considered.

Von Euler and Liljestrand (1946) first demonstrated that making anesthetized cats breathe a hypoxic gas mixture ($F_{O_2}=0.10$–0.11) is invariably associated with an increase of pulmonary arterial pressure. The authors emphasized that "oxygen want and carbon dioxide accumulation ... cause a dilatation of the vessels of the working organs, but call forth a contraction of the lung vessels, thereby increasing the blood flow to better aerated lung areas, which leads to improved conditions for the utilization of the alveolar air" (p 318). Since that time the development of pulmonary hypertension under altitude hypoxia has been repeatedly confirmed in newcomers, sojourners, and native residents at high altitude, as well as in a wide variety of mammals (Hultgren and Grover 1968; Fishman 1976) and in the chicken (Burton et al. 1968; Sillau et al. 1980 b). On the other hand, the extent to which constriction of pulmonary vessels is capable of redistributing blood flow and improving pulmonary gas exchange is poorly documented (Chap. 5.3.2.2). Due to the rise of pulmonary arterial pressure in hypoxia, right ventricular hypertrophy develops, and pulmonary edema sometimes occurs (Hultgren and Marticorena 1978; Lockhart and Saiag 1981).

Striking species differences characterize the reactivity of the pulmonary circulation of man and animals exposed to high altitude. The amount of smooth muscle in pulmonary vasculature inherent within each species appears to be a major determinant of this interspecific variability (Tucker et al. 1975), but cold exposure, diet, unequal ventilatory responses to hypoxia and related differences in alveolar and blood P_{O_2}, variations in responsiveness of pulmonary vascular smooth muscle, and genetically coded predisposition have also been advocated

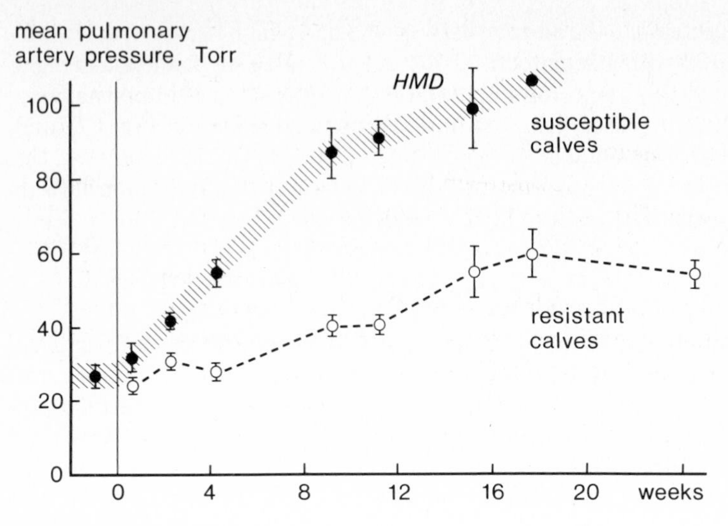

Fig. 4.12. Development of hypoxic pulmonary hypertension (mean values ±SE) and occurrence of high mountain disease (HMD) in calves exposed to the altitude of 3.05 km. (After Will et al. 1975). *Susceptible calves* (n = 8) were born from parents that were not necessarily native to high altitude but had developed HMD at altitudes of 2.4 km or higher. *Resistant calves* (n = 11) were offspring of healthy parents born and raised at altitudes above 2.7 km and were free of HMD. Note the striking difference between the two groups in the development of pulmonary hypertension. None of the resistant calves showed signs of HMD, whereas two of the susceptible animals had developed HMD by 9–10 weeks of exposure, and the remaining six by 18 weeks. The results indicate that susceptibility to pulmonary hypertension was inherited

(Heath and Williams 1981; Lockhart and Saiag 1981). As an example, Fig. 4.12 shows that offspring of cattle with a past history of severe pulmonary hypertension display a disproportionate pulmonary pressure response to high altitude. Right heart failure, due to excessive pulmonary vasoconstriction brought about by hypoxia in the "susceptible" animals, resulted in a clinical syndrome named high mountain disease (HMD in Fig. 4.12; commonly referred to as brisket disease because of the edematous swelling of the brisket). In humans, higher pulmonary arterial pressure is found in patients with chronic mountain sickness, in which severe arterial hypoxemia is attributable to a low hypoxic ventilatory chemoreflex (Lefrançois et al. 1968). However, bovine HMD and human Monge's disease (Sect. 4.2.1.4) differ in many aspects (Heath and Williams 1981).

The polycythemic blood in chronic hypoxia may contribute to pulmonary hypertension (Barer et al. 1983), but the dramatic increase in pulmonary arterial pressure shown in Fig. 4.12 occurred in the absence of change in hematocrit (Will et al. 1975). The cause of hypoxic pulmonary hypertension is vasoconstriction in small vessels, as a consequence of hypoxia. The low P_{O_2} in alveoli and blood can act on the vascular smooth muscle cells, but the actual mechanism is still obscure (Fishman 1976; McMurtry et al. 1982).

Fish have complete branchial blood flow that is identical to the cardiac output (Johansen 1979); its overall changes under hypoxia, therefore, should be also identical (Sect. 4.3.1). That hypoxia affects the regional ventilation–perfusion distribution in gills has been advocated (Holeton and Randall 1967), and perfusion experiments have shown that deoxygenation of perfusion fluid causes gill vascular constriction (Ristori and Laurent 1977). More precisely, Pettersson and Johansen (1982) pointed out that lowering the level of oxygenation increases the branchial vascular resistance, probably due to vasoconstriction proximal to the arterio-venous anastomoses, downstream from the secondary lamellae where gas exchange takes place. They stressed that "vasoconstriction at such sites will, if it affects enough secondary lamellae, cause an increase in afferent lamellar pressure; this response may in turn cause recruitment of unperfused lamellae (Booth 1979)." They also provided evidence suggesting that a combination of a direct myogenic response to hypoxia and release of adrenaline serve to increase O_2 uptake efficiency when fish are exposed to hypoxic stress.

4.4 Summary

The relevant questions here are: How do organisms maintain a sufficient flux of O_2 molecules to match the metabolic demand, when in altitude hypoxia arterial P_{O_2} is lower than normal? How do they spare P_{O_2} along tissue capillaries in order to maintain an otherwise insufficient gradient for O_2 from blood to mitochondria? As stated in Eq. (4.3), the venous P_{O_2} viewed as an index of tissue oxygenation (1) relates directly to arterial P_{O_2} which, as examined in preceding chapters, is dependent upon the anatomic arrangements and functional changes that enhance oxygen uptake from the ambient medium, and (2) relates inversely to the $\dot{M}_{O_2}$-specific blood O_2 conductance ($\beta b_{O_2} \cdot \dot{V}b/\dot{M}_{O_2}$), which itself relates to both the blood O_2 capacitance (βb_{O_2}) and blood flow ($\dot{V}b$).

An increase in the blood O_2 capacitance is a body response to hypoxia common to all vertebrates. The mechanisms by which it is elicited, however, may be different.

First to be considered is the single mechanism unquestionably common to all vertebrates, that which makes use of the shape of the blood O_2 equilibrium curve itself (OEC). With its steep part in the hypoxic range of P_{O_2}, OEC prevents a precipitous drop of the capillary O_2 tension, and increases the blood O_2 capacitance, βb_{O_2}, up to six fold. Instantaneously at work in case of blood O_2 desaturation, this free charge physicochemical process remains operative whatever the duration of hypoxia.

Second, polycythemia is another means of increasing the blood O_2 capacitance under altitude hypoxia, but it is not common to all vertebrates. The process increases both the hemoglobin concentration and, at given P_{O_2}, the blood O_2 concentration. It results essentially from an increased production of erythropoietin, a hormone that stimulates the bone marrow to produce more red cells. There is, however, great intra- and interspecific variability, which is not fully understood, but depends upon many variables such as intensity and duration of hypoxia, associated cold stress, age, sex, nutritional, and genetic factors. Mammals and birds adapted to high altitude for many generations apparently do not rely on increased hemoglobin concentration for adequate O_2 transport to tissues. By contrast, in highlanders as well as in sojourners, and in lowland mammals and birds exposed to high altitude, the increase in hemoglobin concentration does occur. In some individuals, excessive erythropoietic activity is part, if not the cause of the many symptoms of chronic mountain sickness. Since increasing the red-cell mass increases blood viscosity, hence flow resistance and metabolic cost of propelling blood, the "optimal" hematocrit raises questions.

Third, large shifts of the position of the OEC may have functional consequences for O_2 transport by blood, and for tissue oxygenation. Not specific to hypoxic conditions, is the well-known rightward shift that occurs in all vertebrates as blood makes its circuit through the body: through the Bohr effect, this process allows oxygen to be delivered to the tissues at a higher partial pressure. Under altitude hypoxia, depending on the animal species and the duration of hypoxic exposure, both right- and leftward shifts of the OEC have been reported to take place by means of events such as hyperventilatory alkalosis, changes in the ligand phosphate concentration, in the sensitivity of hemoglobin to its allosteric effectors, and in the amino acid sequence of hemoglobin. Neither the magnitude, nor the physiological consequence of these shifts has as yet been adequately assessed. Alterations in Hb-O_2 affinity are generally assumed to be reflected effectively by the changes in a single point of the OEC, namely in the O_2 partial pressure at which 50% of the hemoglobin is saturated with "standard" conditions of pH, P_{CO_2} and temperature, which may be very different from those that prevail in the body. Also, the complicating influences of the age, sex, nutritional state, body hydration, physical fitness, training or lack of training, etc., have not yet been defined. On the other hand, the net influence of wholesale shifts of the OEC depends upon the value at which arterial P_{O_2} is maintained, and the extent to which the differences between arterial and venous O_2 concentration are enlarged across various tissues. Available information suggests that a right shift of the OEC improves the efficiency of the O_2 transport by blood in mild hypoxia, but impairs it in severe hypoxia, whereas a left shift has the opposite effect. There is some evidence that a left-shifted OEC might be a characteristic of the animals living for many generations at high altitude (or in hypoxic waters). Thus, the right shift of the OEC that, in most cases, occurs in sea-level humans and other air-breathers exposed to altitude hypoxia might be an inappropriate response.

Both an increase in cardiac blood flow and its fractional distribution may contribute to the O_2 delivery to the tissues. In mammals and birds, the most readily apparent response to acute hypoxia is tachycardia, probably through stimulation of the sympathico-adrenal system. A decreased cardiac stroke volume, possibly through the local depressant effect of hypoxia and accompanying hypocapnic alkalosis, is also a common feature. Depending on the relative importance of these opposite changes in its components, the cardiac output may increase, or decrease, or remain unchanged. It usually, but not invariably, increases during the early phase of hypoxia, but returns to normal when hypoxic exposure is prolonged. Then, the blood convection requirement ($\dot{V}b/\dot{M}_{O_2}$) remains unaltered by sea-level standards. This holds for humans, other mammals, and birds native to high altitude. Water-breathers appear to develop different patterns for increasing O_2 delivery to tissues. Thus, the generalized circulatory response to hypoxia in fish is an increase in cardiac stroke volume largely offset by a decrease in heart-rate frequency.

Interorgan shifts of blood flow during altitude hypoxia are not yet well defined. Available evidence, however, suggests that regional variations in vascular resistances divert blood flow from regions of low O_2 requirements to supply regions of higher O_2 demand, thereby allowing certain critical tissues, such as heart and working muscles, to function more efficiently during hypoxia.

Among the regional variations in vascular resistances, pulmonary vasoconstriction is a regular feature of man at high altitude, but is also observed in some other mammals and in birds. The associated pulmonary hypertension is a critical element in the pathogenesis of high-altitude pulmonary edema. The physiological significance, as well as the mechanisms of the phenomenon, remain not fully understood.

Chapter 5

Diffusive Processes

Increases in both the ventilatory conductance (Chap. 3) and blood O_2 conductance (Chap. 4) under altitude hypoxia are of adaptive value in raising the partial pressures of O_2 at the gas/blood barrier and in the blood of systemic capillaries. Yet these O_2-driving pressures are lower at high altitude than at sea level (Fig. 2.2). Accordingly, one may wonder how O_2 is transferred satisfactorily across gas/blood and tissue barriers. Do the O_2-diffusive conductances of these barriers increase under altitude hypoxia?

Attempts to answer this question are provided in Sects. 5.3 and 5.5. On the other hand, because the interference of altitude hypoxia with reproduction raises problems (Smith 1973; Clegg 1978) some aspects of gas exchange across the avian eggshell and mammalian placenta are considered in Sects. 5.2 and 5.4. A general approach of the problem is first summarized.

5.1 Basic Considerations

The general Eq. (2.1) states that the gas flux, $\dot{M}_x$, is a direct function of the conductance, G_x, and of the partial pressure difference, ΔP_x, between two locations of the gas-exchange system. G_x may refer to a "complex" conductance when, for anatomical and technical reasons, the two locations under consideration are such that ΔP_x is due not only to the effects of a membrane diffusion resistance, but also to complicating effects, as are those of ventilation-perfusion inequality and/or venous blood admixture in the external gas exchanger (Chap. 2.2.2). Examples will be found when considering O_2 transfer from alveolar gas to arterial blood (Sect. 5.3) or from maternal blood to fetal blood (Sect. 5.4). When ΔP_x is actually determined between the two sides of a diffusion membrane, as across the avian eggshell (Sect. 5.2) or, using indirect methods, from alveolar gas to pulmonary-capillary blood (Sect. 5.3), then G_x refers to a true diffusive conductance. Equation (2.3) states that any diffusive conductance is dependent on (1) the geometry of the membrane, and (2) the physical factors of diffusion.

For the geometry, it is clear that the larger the *functional* surface area (A) and/or the smaller the thickness of the membrane (E) the greater the diffusive conductance. The physical factors of diffusion [the product $D_x \cdot \beta_x$ in Eq. (2.3)], almost certainly do not change much in biological materials if the temperature does not change. Accordingly, the diffusive conductance can be changed only through a change of the membrane geometry. In a gas phase, however, the diffusion coefficient (D_x) increases in direct proportion to the decrease of the barometric pressure (Sect. 2.1.3). This has functional consequence for the avian embryo at high altitude, as examined below.

5.2 Diffusion Across the Eggshell

In avian embryos O_2 transport is a three-step process (1) diffusion through shell and membranes from atmosphere to chorioallantoic capillary blood, (2) convective transport by blood to tissue capillaries, and (3) diffusion from tissue capillaries to cells. Viewing the diffusive gas conductance of the shell as functionally equivalent to the pulmonary ventilation of animals with lungs, some problems related to high altitude are considered below. No attempt is made to discuss all the mechanisms subserving gas exchange in the avian embryo. Comprehensive reviews are available (Freeman and Vince 1974; Scheid 1982). Further discussion about high-altitude adaptation can be found in Rahn (1977, 1983), Rahn et al. (1977), and Carey et al. (1982).

Enclosed in the calcareous eggshell, the avian embryo gains O_2 and loses CO_2 and H_2O (in the form of water vapor) exclusively by diffusion through gas-filled microscopic pores that traverse the shell (Wangensteen and Rahn 1970/71). The number and size of the pores in the eggshell [i.e, the ratio A/E in Eq. (2.3)] is fixed for a given egg at the time of its formation in the parental shell gland; for a chicken egg, as an example, there are about 10,000 pores (3 per 2 mm^2), each 0.3 mm long with an average diameter of 8–12 µm (Wangensteen et al. 1970/71). With a fixed pore geometry, therefore, the avian embryo is critically dependent upon the diffusivity of gas in the pore-enclosed gas phase. Downward the O_2 flux it seems that, in the course of incubation, the embryonic gas exchange organ, namely the chorioallantoic capillary network, develops maximally in normoxic conditions (coming in contact with the whole surface of the inner shell membrane), and that its growth cannot be accelerated by hypoxia (Metcalfe et al. 1979).

5.2.1 Effects of Hypobaria

Rahn and Ar (1974) pointed out that the major problem posed in gas exchange in eggs at high altitude is the inevitable increase of the diffusion coefficient in the gas-filled pores, in direct proportion to the decrease of the barometric pressure ($D_x \propto 760/P_B$). At the altitude of 5.5 km as an example ($P_B = 380$ Torr, or 0.5 atm), the diffusivity of any gas through the eggshell will be twice as great as at sea level. As long as the pore geometry (A/E) remains unaltered, it follows that the diffusive conductance of the eggshell is also twice as great. Consequently, for a given partial pressure difference across the eggshell the gas flux will be twice that found at sea level.

Thus, when fertile eggs from sea level are incubated at high altitude, that is under *hypobaric hypoxia,* enhancement of gas-phase diffusion offers the potential advantage to compensate for the drop of the ambient P_{O_2}. However, as will be seen, it brings with it the threat of hypocapnia and dehydration.

Completely different is *normobaric hypoxia* (lowered O_2 concentration at normal barometric pressure). Visschedijk et al. (1980) clearly showed that, due to the lack of increased gas diffusivity under such a condition, the oxygen uptake is lower than in hypobaric hypoxia at identical ambient P_{O_2}. The following section is concerned exclusively with hypobaric conditions.

5.2.2 Sea-Level Eggs at High Altitude

Erasmus and Rahn (1976) showed that P_{CO_2} in the air cell fell from 32 to 22 Torr when sea-level eggs were exposed to 0.5 atm in a gas mixture containing 40% O_2 in N_2 to prevent hypoxia. Hypobaria per se, and the related increased gas diffusivity were responsible for the observed hypocapnia. Similarly, despite a slightly hyperoxic environment, excessive water loss proved fatal for chicken embryos incubated in a chamber depressurized at about 0.3 atm, unless portions of the eggshell were covered with paraffin to restrict diffusion area (Weiss 1978). Uncoated eggs lost three times as much water as controls at sea level, and they failed to hatch. Water loss and hatchability returned to normal at 69% shell coverage, which balanced the increased diffusion coefficients with the reduced diffusion area.

Due to the enhancement of gas-phase diffusion under hypobaria, and because the magnitude of eggshell conductance is fixed once the egg is laid, fertile eggs from sea level that are transferred to high altitude for incubation require changes in incubator gas composition if they are to develop normally. Equations allow prediction of the necessary enrichment in O_2, CO_2 and water vapor at any altitude (Visschedijk and Rahn 1981). Under the conditions prevailing in mountains (in the range 3.1–3.8 km), reduced metabolic rates, prolonged incubation periods, low body mass at hatching and impaired hatchability have been repeatedly reported for incubated eggs laid at sea level (Smith et al. 1969; Smith 1973; Beattie and Smith 1975). However, with successive generations of hens hatched and maintained at 3.8 km, hatchability was found to improve from the initial rate of 16 vigorous chicks percent of fertile eggs to approximately 60% (Smith 1973). Hence some changes occurred, which are considered below.

5.2.3 Eggs from Birds Breeding at High Altitude

From daily water loss of eggs in a dessicator [$\dot{M}_x$ in Eq. (2.1)] and the difference in water vapor pressure across the eggshell [ΔP_x in Eq. (2.1)], Wangensteen et al. (1974) showed that the water vapor conductance was reduced by a factor of 0.68 in the eggs laid by those hens acclimated for many generations at 3.8 km (Smith 1973), when studied at sea level and compared with their sea-level controls. Since the shell thickness [E in Eq. (2.3)] for both groups of eggs was essentially the same,

the functional pore area (A) must have been reduced in similar p
the shell conductance. At 3.8 km (PB = 480 Torr), such a redu
diffusion surface area counterbalanced rather well the increase in th
of gas, which was 760/480 = 1.58 times greater than at sea level. Acc
gas conductance of the two groups of eggs at their respective altitud
identical.

These results, and similar observations on eggs from a few other lected in the 1.2–3.8 km altitude range (Packard et al. 1977; Rahn et al. 1977) suggest that the total effective pore area of eggshell in a species becomes smaller in proportion to the decrease in barometric pressure at altitudes up to 3.8 km. Whether this reduction is due to a change in the number of pores, or in the pore radius, or in both, is not known. Also unknown are the mechanisms at work in the parental shell gland.

Reduction of pore area of the shell may be viewed as an adaptive change that prevents dehydration at high altitude, and insures survival and hatchability of the embryo. It does not facilitate O_2 uptake from the environment. Thus, when compared with sea-level values, the embryonic O_2 consumption was found to be reduced by a factor of 0.58 in the incubating eggs from those hens acclimated for many generations at 3.8 km (Wangensteen et al. 1974). This reduction in O_2 uptake was made possible by reducing the egg size and, therefore, the embryo size and metabolism, and by prolonging the incubation period. Covering part of the shell of sea-level eggs incubated in air at sea level also results in retarded embryonic growth, with sparing of brain and heart, but with hepatic stunting in excess of body stunting (McCutcheon et al. 1982). The mechanisms of these changes are not understood.

Avian species native to high altitude, on the other hand, appear to have developed mechanisms that support embryonic oxygen consumption, growth, and development (G.K. Snyder et al. 1982a). A reduction in shell conductance undercompensating the increased diffusivity at high altitude, and/or modification of the diffusive resistance of the inner shell membrane may result in a greater O_2 conductance, a smaller ΔP_{O_2} across the shell and its membranes, thereby favoring oxygenation of the embryo (Carey et al. 1982). Improved convective O_2 transport by blood to tissue capillaries, and better diffusion from tissue capillaries to cells may also contribute, as noted in the embryos of the bar headed goose (G.K. Snyder et al. 1981, 1982b).

5.3 Complex Gas/Blood O_2 Transfer

It is widely accepted today that O_2 is transferred across the gas/blood barrier through simple passive diffusion. In the early part of this century, however, there was controversy over whether or not adequate oxygenation of the arterial blood would be possible at high altitude without active secretion of O_2 by the lung (references in West 1981). More recently, it has been advocated by Gurtner (see Wagner 1977, and Knoblauch et al. 1981) that cytochrome P-450 could act as a specific carrier facilitating the diffusion of oxygen.

Diffusion limitation at very high altitude may considerably impair the O_2 transfer across the gas/blood barrier, especially when the O_2 demand is increased under exercise (Sect. 5.3.3). At lower elevations, however, significant impairment of the loading of arterial blood with O_2 may arise through two mechanisms distinct from diffusion which may be referred to collectively as "physiological shunt" (Sect. 5.3.2): (1) uneven distribution of ventilation with respect to perfusion in the various gas exchange units, (2) venous admixture to arterial blood. In mammalian lung, by far the most studied, resistance to diffusion and physiological shunt determine the so-called alveolar-to-arterial P_{O_2} difference which is considered below.

5.3.1 Alveolar-Arterial Difference in P_{O_2}

In mammalian lung, the partial pressure of oxygen in the alveolar gas ($P_{A_{O_2}}$) is higher than in the arterial blood ($P_{a_{O_2}}$; Fig. 2.3). The difference, $P_{A_{O_2}} - P_{a_{O_2}}$, provides an overall appraisal of the O_2 transport across the lung: on the basis of Eq. (2.1), the smaller this P_{O_2} difference, the higher the O_2 conductance between alveoli and arterial blood per unit of oxygen uptake ($\dot{M}_{O_2}$-specific conductance). Therefore, the alveolar-arterial P_{O_2} difference has been considered of great importance in the process of acclimatization to high altitude. Measurements were carried out both in acclimatized lowlanders and in altitude natives (e.g., Cruz et al. 1975; Dempsey et al. 1971; Hurtado 1964; Jaeger et al. 1979) in dogs (Banchero et al. 1976; Kreuzer et al. 1960), and in rats (Turek et al. 1972a).

The most common finding is that the alveolar-arterial P_{O_2} difference is lower, both at rest and during exercise, in native highlanders than in sea-level natives, when studied in their native environments (Fig. 5.1; compare crosses and circles). The only discordant observation is that of Kreuzer et al. (1964), but differences in the technique of measurements or unrecognized pulmonary disease may have

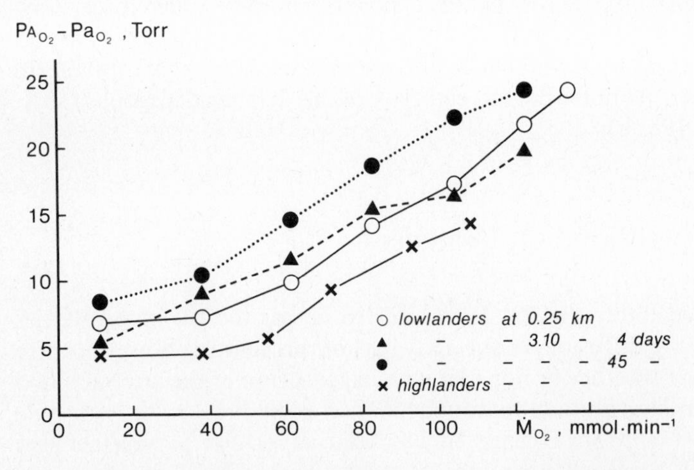

Fig. 5.1. Alveolar-arterial P_{O_2} difference ($P_{A_{O_2}} - P_{a_{O_2}}$) as a function of the O_2 demand ($\dot{M}_{O_2}$) at rest and during exercise near sea level and at high altitude. (After Dempsey et al. 1971)

been responsible (Mithoefer et al. 1972). In growing rats exposed for 30 da the altitude of 3.5 km, the alveolar-arterial P_{O_2} difference was also found lower than in control normoxic animals (Turek et al. 1972a). Thus, improve of the $\dot{M}_{O_2}$-specific conductance across the lung would be brought about by hypoxia associated with the growth of lung.

In newcomers to high altitude, the alveolar-arterial P_{O_2} difference may be differently affected, depending on the severity and duration of hypoxic exposure, and the energy expenditure. Most observations indicate that, in well-adapted subjects, the difference is about the same as in native highlanders (Cerretelli 1980). It failed to decline, however, in those subjects studied by Dempsey et al. (1971) after 4 days at 3.1 km and, moreover, it widened after 45 days (Fig. 5.1, closed circles). A widening of the alveolar-arterial P_{O_2} difference, after arrival at 4.6 km, has been attributed to the development of pulmonary edema (Reeves et al. (1969).

A theoretical study of gas exchange in human lung (Farhi and Rahn 1955) may help in analysing the problem. Figure 5.2 shows that (1) *at given level of oxygenation* (read on the abscissa), the "total" alveolar-arterial P_{O_2} difference results from the additive effects of the prevailing ventilation-perfusion inequality, venous blood admixture, and diffusion limitation at low alveolar P_{O_2}(2) *when going from normoxia to hypoxia* without changing the magnitude of these prevailing factors, the effects of ventilation-perfusion inequality and venous blood admixture on the alveolar-arterial P_{O_2} difference decline, whereas that of diffusion

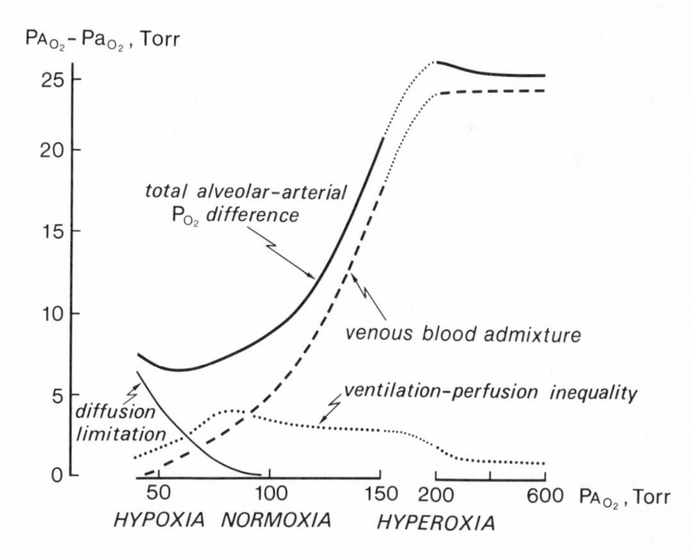

Fig. 5.2. Theoretical effects of changes in the level of oxygenation (*abscissa* $P_{A_{O_2}}$ = alveolar O_2 tension) on the alveolar-arterial P_{O_2} difference ($P_{A_{O_2}}$–Pa_{O_2}) and its components. (After Farhi and Rahn 1955). As an example, at $P_{A_{O_2}}$ = 80 Torr, the total difference is 7.5 Torr, and consequently the mixed arterial blood P_{O_2} is 80–7.5 Torr. In this case, the 7.5 Torr difference between $P_{A_{O_2}}$ and Pa_{O_2} is the sum of the corresponding values read on the component lines, namely diffusion limitation (0.7 Torr), venous blood admixture (2.8 Torr) and ventilation-perfusion inequality (4 Torr). The exact shape of the curves depends on the initial assumptions (see Farhi and Rahn 1955)

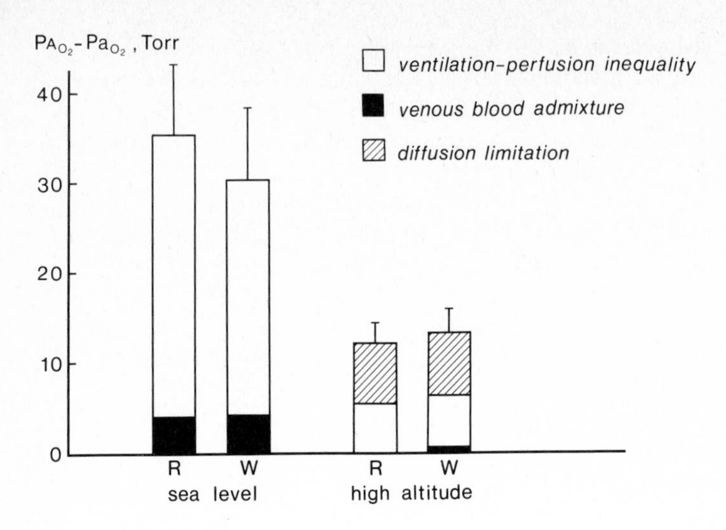

Fig. 5.3. Alveolar-arterial P_{O_2} difference ($PA_{O_2}-Pa_{O_2}$) and its components derived from inert gas data in conscious dogs, either resting (*R*) or exercising on a treadmill (*W*; 8 km·h^{-1}; grade 3–5%) at sea level and at the simulated altitude of 6.1 km in a hypobaric chamber (PB = 349 Torr). (After Sylvester et al. 1981)

limitation increases. The functional consequence is such that, for instance, the alveolar-arterial P_{O_2} difference decreases from 9 to 6.4 Torr when the alveolar P_{O_2} is lowered from 100 to 60 Torr.

The effect of changing the magnitude of any of the components shown in Fig. 5.2 can be intuitively understood. At 60 Torr alveolar P_{O_2}, for instance, reducing the magnitude of the diffusive resistance will decrease the diffusion limitation and will further reduce the alveolar-arterial P_{O_2} difference, thereby improving the O_2 conductance from alveolar gas to arterial blood. Yet this beneficial effect can be overcome by a concomitant increase in the magnitude of either ventilation-perfusion inequality, or venous blood admixture, or both. Due to great technical difficulties, only a few studies have been concerned with these problems at high altitude.

Figure 5.3 shows the results obtained under acute hypoxia in conscious dogs, resting or exercising at the altitude of 6.1 km in a hypobaric chamber (Sylvester et al. 1981). The use of measurements of O_2 and inert gas exchange (see Wagner 1977) allowed the effects of ventilation-perfusion inequality, venous blood admixture and diffusion limitation to be distinguished. Note the large alveolar-arterial P_{O_2} difference in the dogs at sea level, and its important reduction at high altitude. Obviously, the O_2 conductance across the lung per unit oxygen uptake [equal to the reciprocal of the alveolar-arterial P_{O_2} difference, according to Eq. (2.1)] increased considerably under hypoxia. This resulted from improvement in the physiological shunt (the effects of both the ventilation-perfusion inequality and venous blood admixture decreased), despite a concomitant increase in the diffusion limitation. Further examination of these results follows.

5.3.2 Physiological Shunt

5.3.2.1 Venous Admixture to Arterial Blood

In the dog study mentioned above, the magnitude of the venous blood admixture proved to be identical in both normoxia and hypoxia (about 0.6% of the cardiac output). Yet its contribution to the alveolar-arterial P_{O_2} difference became negligible in hypoxia (Fig. 5.3), when the alveolar P_{O_2} was close to 45 Torr. This is in accordance with the analysis shown in Fig. 5.2, and was to be expected according to an important generality: a given small decrease in arterial blood O_2 concentration due to venous contamination is associated with a drop of P_{O_2} in the arterial blood, which is much less in hypoxia than in normoxia; this is because the blood O_2 equilibrium curve is much steeper in hypoxia. By contrast, under hyperoxia (i.e., during inhalation of pure oxygen for a few minutes), the blood O_2 equilibrium is so flat that the same venous contamination provokes a considerable drop of P_{O_2}. Then, as Fig. 5.2 shows, the important alveolar-arterial P_{O_2} difference is essentially due to venous blood admixture (there is no diffusion limitation, and the contribution of ventilation-perfusion inequality is negligible), the magnitude of which can be calculated. Although the oxygen breathing itself may have obscured any difference that exists while breathing air, the method was used in some studies at high altitude. Thus, the magnitude of the venous blood admixture was reported to decrease by about 37% within a few days in lowlanders exposed at 3.8–4.4 km (Kronenberg et al. 1971; Cruz et al. 1975); underlying mechanisms are unknown. Hence, under altitude hypoxia, not only the contribution of the venous blood admixture is less (due to the shape of the blood O_2 equilibrium curve), but its magnitude may become smaller, thereby further improving the lung effectiveness in the process of blood oxygenation.

5.3.2.2 Ventilation-Perfusion Inequality

Not only did the contribution of the ventilation-perfusion inequality decrease in the hypoxic dogs studied by Sylvester et al. (1981, Fig. 5.3), but also its magnitude was reduced. Two explanations can be advanced: (1) the overall mean ventilation/perfusion ratio increased, because the alveolar ventilation was 60% higher at 6.1 km altitude than at sea level, whereas the cardiac output remained virtually unaltered, and (2) the regional ventilation/perfusion ratios throughout the lung became more homogeneously distributed, concomitantly with a 50% increase in pulmonary arterial pressure. A similar observation was made in humans exposed to 3.1 km. Using radioactive xenon and collimated scintillation detectors, Dawson and Grover (1974) found the relative ventilation-perfusion to be more uniformly distributed from apex to base in natives and long-term residents than in normal subjects at sea level.

By contrast, Haab et al. (1969) found the ventilation-perfusion inequality to be increased during the first 5 days of a stay at 3.5 km altitude. They used a method based on the analysis of alveolar-arterial P_{CO_2} and P_{N_2} differences. Neither one of these was modified by ascent, and there was no significant change during the stay at high altitude. Had the ventilation-perfusion distribution remained unchanged, the decrease in ambient pressure would have substantially lowered the alveolar-arterial P_{N_2} difference. The authors concluded that the relocation from low to high altitude altered the scatter of ventilation or perfusion, or both, in such a way as to decrease the gas-exchange efficiency of the lung; this increase in ventilation-perfusion inequality offset partly the benefits derived from the hyperventilation response. Other indirect evidence also supports the view that ventilation-perfusion inequality would increase rather than decrease in highlanders (Kreuzer et al. 1964; Mithoefer et al. 1972; Saltin et al. 1968).

There is no clear explanation for these apparently opposing findings. Perhaps wide-spread ventilation-perfusion abnormalities occur at the alveolar level despite better regional matching of ventilation and blood flow. The occurrence of micropulmonary thromboemboli due to increased hematocrit and blood viscosity, as well as diffuse, asymptomatic pulmonary edema have been advocated (Lenfant and Sullivan 1971).

Interestingly, calculations based on a computer model (West 1969) have shown that the maximal influence of ventilation-perfusion inequality on the alveolar-arterial P_{O_2} difference (peak of the dotted curve in Fig. 5.2) is shifted toward higher alveolar P_{O_2} values when the ventilation-perfusion inequality increases. This phenomenon may explain why, despite larger ventilation-perfusion inequality in some highlanders, these subjects may have in their hypoxic environment the same alveolar-arterial P_{O_2} difference as sea-level residents breathing air or hypoxic gas mixtures (Mithoefer et al. 1972).

5.3.3 Diffusing Capacity

When the alveolar O_2 partial pressure falls below 80 Torr with ascent to high altitudes, diffusion limitation will theoretically begin to impose a significant "bottleneck" for O_2 transfer across the pulmonary respiratory surface (Fig. 5.2). Thus any change in the gas/blood barrier yielding reduction of this diffusion limitation may be an important adaptive mechanism.

5.3.3.1 Methodological Approach

Classically, the total resistance to diffusion between pulmonary gas and blood (1/GL) results from three resistances in series. This can be written

$$\frac{1}{G_L} = \frac{1}{G_{gas}} + \frac{1}{G_M} + \frac{1}{G_b}, \tag{5.1}$$

where 1/Ggas is the gas-phase resistance due to incomplete mixing or stratification is small conducting airways, 1/GM is the pulmonary membrane resistance, which depends primarily on thickness and surface area, and 1/Gb is a resistive kinetic component which depends on the transit time of blood in the pulmonary capillaries and the rate of reaction between oxygen and hemoglobin. Similar reasoning holds for nonmammalian vertebrates [for water-breathers, 1/Gwater would simply replace 1/Ggas in Eq. (5.1)]. Any reduction in these serial resistances will increase the pulmonary O_2 conductance (GL, often named pulmonary or lung-diffusing capacity), thereby enhancing the O_2 transfer form pulmonary gas to hemoglobin-binding sites in red cells.

To date, the degree of limitation that stratification exerts on gas exchange has not been clarified. Some evidence indicates that the stratificational resistance might be 10–20% of the total gas/blood resistance to O_2 transfer in air breathers near sea level (Scheid and Piiper 1980). Presumably, this contribution decreases under hypobaric hypoxia, since the diffusivity of gas increases as an inverse function of barometric pressure.

In all the studies so far carried out, GL, the measured diffusing capacity, was supposed to comprise the conductance across the stratified gas phase in small conducting airways, Ggas. Therefore, Eq. (5.1) becomes $1/G_L = 1/G_M + 1/Gb$, where $1/G_M$ is the membrane resistance, and $1/Gb$ the intracapillary resistance from plasma to hemoglobin-binding sites in red cells. Attempts have been made to analyze the effects of altitude hypoxia on these two components of the diffusion barrier. The underlying rationale is as follows.

Gb is the diffusing capacity of the red cells in the pulmonary capillary bed at any moment. It is identical to the product $\theta \cdot Vc$ in the classical equation derived by Roughton and Forster (1957), where Vc is the capillary blood volume undergoing gas exchange, and θ stands for the amount of oxygen which can be bound by 1 ml of blood in 1 min under 1 Torr partial pressure, and therefore is the diffusing capacity of 1 ml of whole blood. Accordingly, the above equation can be rewritten $1/G_L = 1/G_M + 1/\theta \cdot Vc$.

The value of θ, as determined in vitro, has been found to increase when the blood O_2 saturation decreases (see Wagner 1977). By making measurements of the overall lung-diffusing capacity, GL, at different levels of oxygenation (altitudes), thereby causing θ to vary, the latter equation can be solved graphically for GM, the membrane-diffusing capacity, and Vc: the plot of $1/G_L$ against $1/\theta$ theoretically yields a straight line with y-axis intercept at $1/G_M$ and a slope of $1/Vc$.

There are uncertainties, however. Assumedly not affected by altitude hypoxia, the values of θ are taken from the literature and corrected for the increased blood hemoglobin concentration. On the other hand, if GL may be determined theoretically on the basis of Eq. (2.2), the partial pressure difference to be determined is that between alveolar gas and average pulmonary capillary blood. The latter being inaccessible to direct measurement, indirect methods are necessary. In virtually all studies, subjects are given about 0.2% CO in air to breathe; the affinity of hemoglobin for CO is such that the mean pulmonary partial pressure of CO can be neglected, and accordingly the diffusing capacity for CO is simply calculated as $G_{L_{CO}} = \dot{M}_{CO}/P_{A_{CO}}$ on the basis of Eq. (2.2). Yet the relationship between the diffusing capacity for CO and that for O_2 is poorly understood (see Piiper and Scheid 1980). Consequently, the effects of altitude hypoxia on the lung O_2-diffusing capacity remains an area of uncertainty.

5.3.3.2 Acute and Prolonged Hypoxia

In sojourners no consistent changes in the lung-diffusing capacity for carbon monoxide ($G_{L_{CO}}$) have been observed in the course of exposures up to 3 months at altitudes up to 4.7 km (Dempsey et al. 1971; Guleria et al. 1971; Kreuzer and Van Lookeren Campagne 1965; West 1969). The $1/GL_{CO}$–$1/\theta_{CO}$ relationships were found to increase slightly in slope (West 1969) or to decrease (Dempsey 1971), with an unchanging intercept; hence, the capillary blood volume exhibited variable changes, but the membrane-diffusing capacity was unaltered. During exercise, GL_{CO} did increase, but the observed changes were the same as at sea level under similar work loads. Only at 5.8 km, has a significant 15–20% increase above sea-level values been reported by West (1962) in exercising acclimatized subjects. However, this small increase in GL_{CO} could be wholly explained by the more rapid reaction of carbon monoxide with hemoglobin caused by the low blood O_2 saturation and increased hemoglobin concentration. The same was found in growing rats adapted for 23–30 days at the altitude of 3.5 km in a hypobaric chamber (Turek et al. 1972a).

Although scanty, these observations suggest that sea-level residents do not have the capacity to increase their lung-diffusing capacity within a few months' period of acclimatization at high altitude. This can be detrimental when the O_2 needs are increased, as during physical exercise. Thus, in lowlanders acclimatized to 5.8 km, West et al. (1962) noted a marked decrease in the arterial blood O_2 saturation during exercise. This fall occurred despite a simultaneous increase in alveolar P_{O_2} due to relative hyperventilation; as a result the difference between the alveolar and arterial P_{O_2} increased, suggesting diffusion limitation in the lung.

Such a deleterious effect of hypoxia on diffusion between pulmonary gas and blood is illustrated in Fig. 5.4, which shows both the results of theoretical calculations (West and Wagner 1980) and some data obtained on the recent American Medical Research Expedition to Everest led by West (1983). In panel A, note the relentless decrease of P_{O_2} in arterial and mixed venous blood as the O_2 demand is increased. Panel B shows how the value of P_{O_2} in mixed venous blood falls with increasing O_2 demand for different values of the pulmonary membrane diffusing capacity, GM_{O_2}. Assuming that further O_2 utilization ceases when the mixed venous P_{O_2} falls below the critical value of 15 Torr (dashed line in Fig. 5.4B), it is clear that the lower the membrane-diffusing capacity, the lower the permissible maximal oxygen uptake, $\dot{M}_{O_2}^{max}$. This deterioration of $\dot{M}_{O_2}^{max}$ with decreasing GM_{O_2} is illustrated in Fig. 5.4C.

Why is O_2 transfer so diffusion-limited under these conditions? Figure 5.4D shows the result of a calculation of oxygenation along the pulmonary capillary for a climber on the summit of Mt. Everest with a membrane diffusing capacity of 4.5 $mmol \cdot min^{-1} \cdot Torr^{-1}$. Whereas near sea level full equilibration of capillary and alveolar P_{O_2} would have occurred within about 0.3 sec, note the slow increase of P_{O_2} along the capillary and the 7 Torr P_{O_2} difference between alveolar gas and end-capillary blood after 0.75 sec. This emphasizes the marked diffusion limitation to O_2 transfer under these conditions.

The chief reason that O_2 transfer is so limited under hypoxia is the large increase of the blood O_2 capacitance coefficient (βb_{O_2}). Indeed, the ratio of tissue to blood solubility (capacitance) is a major determinant of diffusion rate (see Wagner 1977); under altitude hypoxia, this ratio decreases because the O_2 capacitance of the pulmonary membrane presumably does not change, whereas that of blood increases four- to fivefold (Chap. 4.2). Similarly, Piiper and Scheid (1980)

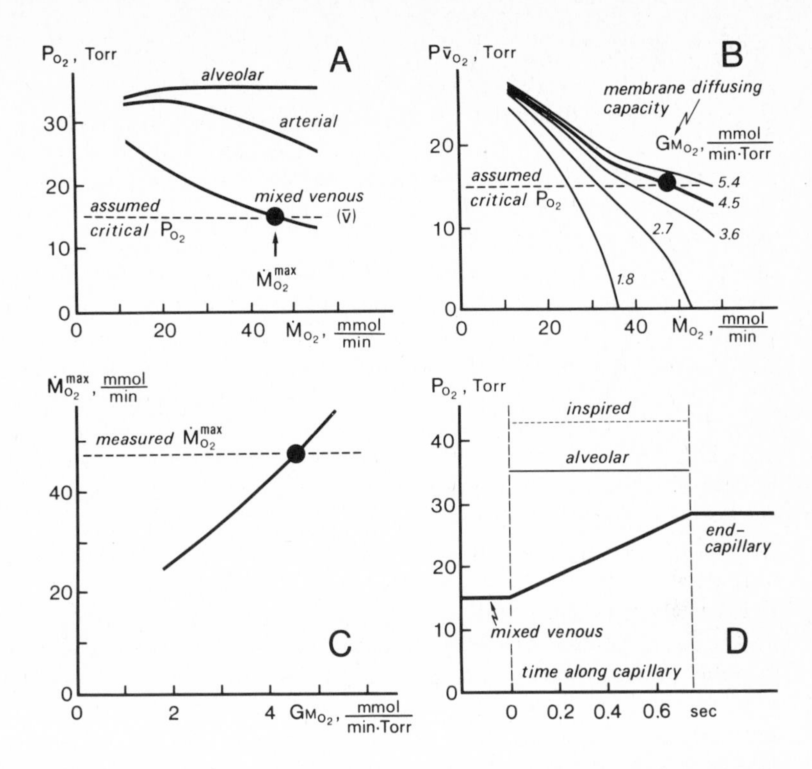

Fig. 5.4. Alveolar-capillary O_2 transfer and some other aspects of gas exchange during maximal exercise at extreme altitude. (After West 1983). *A* theoretical changes in the O_2 partial pressure (P_{O_2}) of alveolar gas, arterial blood and mixed venous blood as oxygen uptake ($\dot{M}_{O_2}$) is increased for a climber breathing air on the summit of Mt. Everest, assuming a constant membrane O_2-diffusing capacity of 4.5 mmol·min^{-1}·Torr^{-1}. Maximal oxygen uptake ($\dot{M}_{O_2}^{max}$) was measured at the altitude of 6.3 km by making the subject to breathe a hypoxic gas mixture simulating conditions on the summit itself. The value of P_{O_2} in the mixed venous blood was assumed to be 15 Torr when $\dot{M}_{O_2}^{max}$ occurred (*dot*). *B* theoretical effects of changes in membrane O_2-diffusing capacity ($G_{M_{O_2}}$) on the O_2 partial pressure of mixed venous blood ($P\bar{v}_{O_2}$) as a function of oxygen uptake for man on the summit of Mt. Everest. Note that a value of 4.5 mmol·min^{-1}·Torr^{-1} for $G_{M_{O_2}}$ is required for the observed $\dot{M}_{O_2}^{max}$ (*dot*). *C* predicted maximal O_2 uptake against membrane O_2 diffusing capacity for man on the summit of Mt. Everest; derived from *B*, assuming the various values of $\dot{M}_{O_2}^{max}$ are all reached when P_{O_2} in the mixed venous blood is 15 Torr. *D* calculated time course of P_{O_2} changes along the pulmonary capillary for man exercising maximally on the summit of Mt. Everest; the membrane O_2-diffusing capacity is assumed to be 4.5 mmol·min^{-1}·Torr^{-1} (*dot* in other panels), and the capillary transit time to be normal (0.75 s). Note the slow increase of P_{O_2} along the capillary, and the P_{O_2} difference between alveolar gas and end-capillary blood

have shown that diffusion limitation occurs when the ratio $G_{M_{O_2}}/\dot{V}b \cdot \beta b_{O_2}$ decreases ($\dot{V}b$, pulmonary blood flow). They have calculated that, under heavy exercise at the altitude of 5.35 km, the reduction in this ratio is such that the O_2 transfer across the lung is reduced by 36%, due to diffusion limitation despite a "normal" diffusing capacity (Piiper and Scheid 1981). This theoretical finding is well in agreement with the reduction of the maximal O_2 uptake observed at that altitude (Fig. 2.5).

5.3.3.3 Chronic Hypoxia

In highlanders, either natives or long-term residents (1–16 years), resting or exercising in the 3.1–4.5 km altitude range, the pulmonary diffusing capacity for carbon monoxide ($G_{L_{CO}}$) has been consistently reported to be 20–30% higher than the values predicted for sea-level conditions or found in newcomers (DeGraff et al. 1970; Dempsey et al. 1971; Guleria et al. 1971; Remmers and Mithoefer 1969). The $1/G_{L_{CO}} - 1/\theta_{CO}$ relationships generally showed a decrease in both the slope and y axis intercept, hence suggesting that both the capillary blood volume, Vc, and membrane-diffusing capacity (conductance), G_M, were elevated. In other words, these studies indicate that both the pulmonary membrane resistance and intracapillary resistance from plasma to hemoglobin-binding sites in red cells would be less in highlanders than in lowlanders. Underlying mechanisms are unclear, but various explanations have been advanced.

The augmented Vc has been ascribed to the increase in total blood volume related to hypoxic polycythemia, with "... lung capillaries sharing in the generalized increase in vascular volume ..." (DeGraff et al. 1970, p 75). Change in the distribution of total blood volume leading to an increased central fraction, and opening of new pulmonary capillaries have been also advocated (Lenfant and Sullivan 1971). That hypoxia causes pulmonary capillary recruitment and raises diffusing capacity for carbon monoxide has been reported by Capen and Wagner in several studies (e.g., 1982) on anesthetized dogs with implanted thoracic windows. However, not hypocapnic but normocapnic hypoxia was studied. Whatever the mechanisms are, an increase in capillary volume, Vc, has the same effect as lengthening the capillary red cell transit time, $Vc/\dot{V}b$, since the cardiac blood flow, $\dot{V}b$, is not increased in chronic hypoxia (Chap. 4.3.1). This condition may help in reducing the P_{O_2} difference between alveolar gas and end-capillary blood (Fig. 5.4D), thereby conferring a serious advantage to the highlanders over the sojourners.

Although no information is available in high-altitude animal species, it is worth noting that, apart from Vc, the diffusing capacity of 1 ml of whole blood, θ (Sect. 5.3.3.1), also determines the intracapillary resistance to gas transfer ($1/\theta \cdot Vc$). As for O_2, θ_{O_2} can be viewed as proportional to the velocity constant for the reaction of O_2 with deoxyhemoglobin in whole blood, k'c (Staub et al. 1962); the greater k'c, the greater θ_{O_2}, and the lower the intracapillary resistance to O_2 uptake. Figure 5.5 suggests that this resistance would be lessened in the case of animals such as the goat and the sheep, and perhaps the llama and the vicuña, whose red cells have a higher k'c value. In the goat, for instance, the resistance would be approximately half that in man. Drawn from the data of Holland and Forster (1966), Fig. 5.5 also indicates that k'c inversely relates to the red cell volume. The differences in resistance, therefore, may represent a true effect of size, conferring some biological advantage on the species with small red cells: the smaller the red cell volume, the shorter the distance for O_2 molecules to diffuse, and consequently the higher the oxygenation rate. On the other hand, the high-altitude-induced polycythemia may decrease the endothelium to red cell distance, thereby shortening the intracapillary diffusion length. Whether this advantage may be significant at high altitude has not been studied. Interestingly, however, in the aquatic frogs of the high Andean lakes, the erythrocyte volume

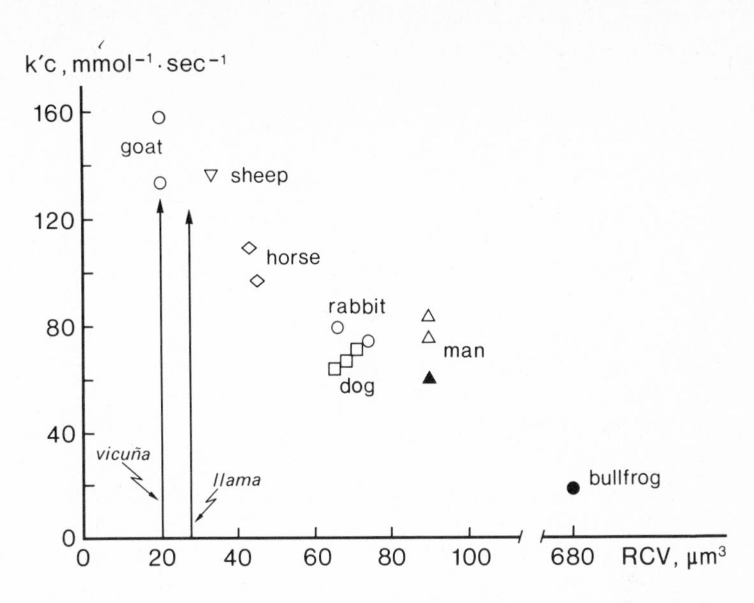

Fig. 5.5. Velocity constant of red cell oxygenation (k′c) plotted against mean red cell volume (RCV). (Inspired by Monge and Whittembury 1976). Plotting and data for goat, sheep, horse, rabbit, man, and bullfrog are from Holland and Forster (1966), at pH = 7.4, P_{CO_2} = 40 Torr, and 37 °C (*open symbols*) or 27 °C (*closed symbols*). *Arrows* visualize the RCV values for llama and vicuña reported by Hall et al. (1936), and the corresponding expected k′c values; they suggest that high-altitude camelids share with sheep and goats the advantage of a higher oxygenation rate

is the smallest known among amphibians (394 μm^3), roughly half that of the bullfrog's red cells shown in Fig. 5.5 (Hutchison et al. 1976).

A greater membrane-diffusive conductance, GM, in residents at high altitude suggests adaptation by the lung to life under chronic hypoxia. Its underlying mechanisms, however, remain an enigma. Theoretically, the changes in GM should follow those in Vc, since the *functional* membrane area is the area of membrane over the perfused capillaries, not the entire alveolar surface area. As discussed by DeGraff et al. (1970), the factors, apart from an expansion of the capillary bed, that may theoretically affect GM are numerous: e.g., simple stretch of lung tissue due to expanded functional residual capacity, growth of new lung tissue, increased numbers of alveoli. Some of these are examined below.

5.3.3.4 Morphological Studies

Although the basic architecture of the lung is established during fetal life, the story of the development of the alveoli is postnatal (Reid 1967). If O_2 needs are augmented during the growth phase of an animal, as by exercise or exposure to cold, the lung appears to be able to adapt quantitatively by increasing primarily the alveolar surface area and capillary volume (Weibel 1979). Whether comparable structural adaptation occurs in animals made chronically hypoxic has been investigated in a limited number of studies.

Table 5.1. Percent changes from normoxic controls in body mass, lung volume, and morphometrical data of lung parenchyma, induced by a few weeks' exposure at high altitude

Animal species		Rat			Guinea pig	
References		1	2	3	4	4
Experimental conditions						
P_{O_2} in dry gas	Torr	78	95	104	83	83
(Equivalent) altitude	km	5.70	4.20	3.45	5.10	5.10
Ambient temperature	°C	25	24	23.5	22	22
Duration of exposure	weeks	2	3	3	2	14
Age at sacrifice	weeks	6	7	6	5	17
Percent changes from control						
Body mass		−19	−7	−11	NS	NS
Lung volume	VL	NS	27	8	39	NS
Alveolar surface area	Sa	NS	25	17	32	NS
Capillary surface area	Sc	NS	25	17	32	NS
Morphometric diffusing capacity	GL_{O_2}			20		

References: 1. Bartlett (1970); 2. Bartlett and Remmers (1971); 3. Burri and Weibel (1971 a); 4. Lechner and Banchero (1980)
(NS = not significant)

Table 5.1 indicates that although hypoxia has an inhibitory effect on body growth, at least in rats, the lung seems not be affected. On the contrary, the lung volume (measured by water displacement), as well as the alveolar and capillary surface area (determined by microscopic morphometry) per unit body mass, are augmented in young, growing animals made hypoxic for a few weeks, as compared with controls. These changes are achieved by an augmentation in both the number and size of alveoli (Burri and Weibel 1971 a; Cunningham et al. 1974).

In guinea pigs, Lechner and Banchero (1980) found that the difference between hypoxic and normoxic animals was progressively reduced with increasing time of hypoxic exposure (Table 5.1; compare the two last columns); apparent cessation of a hypoxia-induced effect occurred despite continued hypoxic exposure for 14 weeks, when the animals were of adult body mass and about 17 weeks of age. This observation indicates that, in guinea pigs, the lung development in chronic hypoxia can be accelerated toward normal adult dimensions, but is ultimately constrained. Cunningham et al. (1974) also noted in rats that exposure to hypoxia for 6 rather than 3 weeks did not produce aditional lung growth. A possible critical factor was that the maximum size of the thoracic cage was then attained, with no available thoracic space for further pulmonary development. This might explain why Tenney and Remmers (1966) found no significant differences in weight-specific lung volume (VL) and alveolar surface area (Sa) in adult guinea pigs raised for several generations at 4.53 km elevation in Peru, compared to sea levels animals. Yet, the mean value for Sa was 18% higher in the high-altitude group. As noted by Burri and Weibel (1971 a, p 261) "it may be that with a larger sample and whith a more refined morphometric technique significance could have been reached." Lechner (1977) also found that the variation associated with a mean increase of 14% in VL and of 22% in Sa did not make fossorial gophers, acclimated for 3 months to simulated high altitude (3.4 km), significantly different from sea-level controls.

Apart from the difficulty in analyzing data obtained from wild animals, for which information on age and development cannot be obtained, the type, duration, and severity of hypoxia have to be considered. Thus, in Bartlett's study (1970, Table 5.1, column 1), exposure of growing rats for 2 weeks to normobaric hypoxia equivalent to 5.7 km altitude slowed somatic growth, and resulted in virtually no change in lung development. By contrast, when animals of identical strain, age and body mass were exposed for 3 weeks to hypobaric hypoxia in a chamber depressurized at 4.2 km, small reduction in body mass but significant increases in both VL and Sa were observed (Bartlett and Remmers 1971, Table 5.1, column 2). In this second study, some animals were killed after 7 days of hypoxic exposure. They had lungs that were abnormally heavy relative to their volume. This increase in lung density was due to an increase in nonblood lung water, presumably as edema fluid. In the animals killed after 3 weeks, however, lung density and water content had returned essentially to normal. In the authors' opinion, if the exposure in the earlier study of Bartlett had been continued somewhat longer, the edema also observed would have subsided, and an increased rate of lung growth would have ensued.

Two comprehensive morphometric analyses have been carried out by Burri and Weibel (1971 a, b). In a short-term experiment, young rats in their most active growth phase (3 weeks old) were exposed for 3 weeks at the altitude of 3.45 km in the Bernese Alps. The lung volume and alveolar surface area per unit body mass increased in the hypoxic animals compared with controls maintained in Berne, and the specific capillary surface area changed in strict proportion to specific alveolar surface (Table 5.1, column 3). Using a model previously described (Weibel 1973), the pulmonary diffusive conductance for oxygen (GL_{O_2} in Table 5.1) was calculated from the morphometric data, which included the mean thickness of the gas/blood barrier; in the hypoxic animals, the thickness did not change, but the conductance increased by 20%. In a long-term experiment, pregnant rats were exposed to the same altitude, 10 days prior to delivery; they were maintained at that altitude with their offspring for about 19 weeks (Burri and Weibel 1971 b). In these growing animals, which were raised at high altitude from birth onward, the changes in body mass, specific lung volume and specific alveolar surface area were very similar to those observed in the previous experiment. By contrast, the mothers of these animals, which were brought to high altitude as adults and stayed there as long as their offspring, showed absolutely no difference from the mothers of the controls, either in body mass or in lung volume.

Altogether these observations indicate that, under prolonged hypoxia, the growth of the lung with respect to body growth can be accelerated toward normal adult dimensions. This appears not to be a specific response, since similar changes are observed upon exposure either to hypoxia alone, or to simple cold, or to associated cold and hypoxia (Lechner et al. 1982). On the other hand, adult animals appear to be no longer capable of adapting their lung structure in response to hypoxia.

It seems that no morphometric study of the lung of high-altitude natives has been conducted yet. Thus no morphological evidence is at present available to explain the increase in diffusing capacity of pulmonary membrane reported in native highlanders.

5.3.3.5 Indirect Evidence

Although no direct evidence from postmortem morphometric studies is at present available to support the idea that the alveolar surface area is increased in those highlanders who take benefit of an above-normal pulmonary diffusive conductance (Sect. 5.3.3.3), there are clinical and functional observations suggesting greater alveolar surface at high altitude. Proof of alveolar distension is lacking, but hyperinflation of the lung is documented.

A barrel-shaped chest cage has long been regarded as characteristic of the high-altitude populations in the Andes of South America (ref. in Heath and Williams 1981). This, as recently examined by Heath (1982), is not a characteristic of all highland populations, but *in general* highlanders tend to have large chests, in comparison with lowlanders of similar stature. The volume of gas remaining in the lung at the end of a normal expiration (functional residual capacity, FRC), and that remaining after a forced expiration (residual volume, V_L) have been reported to be 17% and 38% greater in native residents at 4.54 km than in sea-level natives (Hurtado 1964). Such a hyperinflation of the lung might be one factor in increasing pulmonary diffusing capacity. DeGraff et al. (1970) found both the lung volume and diffusing capacity for carbon monoxide to be greater in life-long residents at 3.1 km elevation than in their sea-level counterparts. It seems that the increased lung volume proceeds from enlargement of the alveolar spaces, but not of the airways. Brody et al. (1977) observed that maximal gas flow rates (expelled during a forced expiration) were less in highlanders than in lowlanders, when expressed as a function of the lung volume at given transpulmonary pressure. This indicates that in the highlanders the cross-sectional area or volume of the airways do not share in the lung enlargement. They would have done so, in these authors opinion, if the growth of lung had been associated with genetic or early fetal adaptation. Therefore, the large lungs of highlanders may result from postnatal environmental hypoxic stimulation of lung growth.

If the lung diffusing capacity does not seem to increase in lowlanders translocated to high altitudes (Sect. 5.3.3.2), is there increase in lung volume? Barcroft et al. (1923) found no change in V_L in his colleagues after 2 weeks at about 4.3 km altitude in the Andes, but Tenney et al. (1953) described a 10% increase in both FRC and V_L in 3 over 4 sea level natives after a 7-day stay at a similar altitude in the Rocky Mountains. As for animals, Barer et al. (1978) found FRC to be much larger in rats that had been acclimatized for 3 weeks to normobaric hypoxia simulating an altitude of 4.3 km. Even only a few minutes of normobaric hypoxia (equivalent to 5 km elevation) made FRC rise by 5–22% in 40 of 43 subjects studied by Garfinkel and Fitzgerald (1978), but Kellogg and Mines (1975) found no change under similar conditions. If human data are conflicting, FRC has been repeatedly reported to increase in rats, rabbits, cats, and dogs made acutely hypoxic (see Barer et al. 1978). As mentioned in Chap. 3.2.1.1 a, this phenomenon may result from an inspiratory shift of neural and muscular activities, and may be due to carotid body stimulation by hypoxia. Functional significance, however, awaits elucidation.

5.3.4 Comparative Data

Very little is known about the adjustments of the O_2 diffusive conductance to altitude hypoxia in nonmammalian vertebrates.

For birds, Lutz and Schmidt-Nielsen (1977) found the O_2 partial pressure difference between end-tidal gas and arterial blood to be large in normoxic pigeons, and to decrease under acute exposure to altitudes up to 9.15 km in a hypobaric chamber. In normoxia the large difference observed means that the arterial P_{O_2} was lower than it could be predicted for an ideal crosscurrent system (Chap. 2.2.1 b); the quantitative influences of ventilation-perfusion inequality, venous blood admixture, and diffusion limitation remain to be clarified (Scheid 1979; Powell 1981). In hypoxia, the smaller P_{O_2} difference between end-tidal gas and arterial blood indicates greater efficacy of the O_2 transfer across the lung; it may be due to both a larger blood O_2 capacitance (Scheid 1979) and a reduced ventilation-perfusion inequality (Escobedo et al. 1978).

In fish, exposure to hypoxic water induces circulatory responses (Chap. 4.3.2.6) that yield perfusion of unperfused lamellae (recruitment). As a functional consequence, reduced ventilation-perfusion inequality, and increased diffusing capacity of gills, due to both reduction of the width of the interlamellar spaces and increased functional surface area, would facilitate oxygen uptake. Recently reviewed by Johansen (1982), these processes are not well quantified yet.

As for amphibians, morphological adaptations allowing an aquatic life at high altitude have been described. Thus, in the frogs of Lake Titicaca (altitude = 3.81 km), the specialized skin, which serves as a gill, is richly vascularized with an increased surface area to body volume ratio provided by a dorsoventrally flattened body and reticulated folds of skin (Hutchison et al. 1976). In an early report, Bond (1960) showed that enlargement of the gills, especially of the filaments, occurs in salamanders when subjected to surroundings deficient in oxygen. More recently, larval bullfrogs exposed for 4 weeks to hypoxic water ($P_{O_2}=70$ Torr) were found to present marked gill hypertrophy, increased lung volume, and greater skin capillarization (Burggren and Mwalukoma 1983). However, adult bullfrogs exposed to the same hypoxic conditions showed no significant changes in the morphology of the skin or lungs; instead adjustments, such as polycythemia and increased hemoglobin concentration, occurred in the blood (Pinder and Burggren 1983). What emerges, then, are quite different responses to long-term exposure to low oxygen levels by different developmental stages of bullfrogs. Quoting the authors' conclusion, "... whether the cost of changes in the morphology of respiration structures vs. changes in blood properties and physiological responses is much greater in the terminal adult stage compared to the tadpole stages, or whether the adult simply lacks regulatory mechanisms for tissue growth which are stimulated by environmental hypoxia, remains to be demonstrated" (Burggren and Mwalukoma 1983).

5.4 Maternal–Fetal O_2 Transfer

The chain of oxygen supply to the developing fetus in utero suffers the disadvantage inherent in complexity (Metcalfe et al. 1967; Longo and Bartels 1972). Because of the various resistances that oppose to the O_2 transfer from the maternal erythrocyte to the fetal erythrocyte, the O_2 partial pressure in the arterialized fetal blood, in the umbilical vein, barely exceeds 20 Torr in normoxic conditions. Hence, even at sea level the fetus is severely hypoxemic. In the adult, such a low arterial P_{O_2} would be found at the summit of Mt. Everest or above. This being so, "... one might anticipate that during pregnancy at high altitude, where oxygen tension in the systemic arteries and hence maternal placental capillaries is low, even more pronounced hypoxemia might be suffered by the fetus, perhaps endangering its very survival" (Heath and Williams 1981, p 263).

However, Metcalfe et al. (1962) reported a remarkable fact from comparison of pregnant ewes bred and pastured at 4.54 km in the Peruvian Andes with individuals kept near sea level. They found that, despite a reduction by half of the arterial P_{O_2} in the high-altitude mothers compared to sea-level ones, the O_2 tensions in the fetal blood entering and leaving the placenta were the same in the two groups. They also noted that the fetuses at high altitude grew at the same rate as those at sea level, and were not meeting any detectable fraction of their energy requirements by anaerobic metabolism. Hence, in the high-altitude group, adaptive changes occurred in the complex O_2 transfer from maternal blood to fetal blood. Decreases in the proportion of the blood shunted away from the placental exchange vessels (Metcalfe et al. 1967) and/or a more efficient balance of the maternal and fetal perfusion rates may have been involved (Power et al. 1967).

In an effort to understand some of the maternal and fetal adaptations to chronic hypoxia, Gilbert et al. (1979) measured the overall placental diffusing capacity for carbon monoxide, Gp_{CO}, in pregnant guinea pigs either kept in normoxic conditions or exposed to normobaric hypoxia (12% O_2; equivalent altitude = 4.5 km) from 18–22 days gestation to the end of pregnancy (62–64 days gestation). When expressed per unit body mass of the fetuses, Gp_{CO} was found to be increased by 63% in the hypoxic animals compared with controls. The placental weights of both groups were the same, but the body mass of the hypoxic fetuses was reduced by about 40%; consequently the placental-to-fetal body mass ratio was increased in the same proportion in the hypoxic animals. The increased Gp_{CO}, however, did not simply result from the decreased fetal weight; the absolute value of the placental diffusing capacity, as well as its value normalized to placental weight, also increased.

The increase in placental diffusing capacity could be explained by an increase in the surface area for gas exchange [A in Eq. (2.3)]. This has been advocated, but not documented, by Barron et al. (1964) on the basis of their observations on the oxygenation of fetal llamas (Meschia et al. 1960). Another possible explanation is a reduction in the diffusion distance between maternal and fetal blood [E in Eq. (2.3)]. Tominaga and Page (1966; quoted by Clegg 1978) artificially perfused fresh human placentae with hypoxic blood (equilibrated with 6% oxygen). They noted a thinning of the maternal–fetal barrier, which was quite obvious after 6 h perfusion, potentially increasing the diffusing capacity by 25%, and which reversed when the level of oxygenation returned to normal. Theoretical studies (Longo et al. 1972) indicate that maternal polycythemia and increased capillary blood volume may contribute to increase the diffusing capacity. Microsomal cytochrome P-450 has been implicated as a possible carrier that would facilitate oxygen diffusion across the placenta (Gurtner and Burns 1972). However, the concentration of cytochrome P-450 did not increase in the hypoxic animals studied by Gilbert et al. (1979); this carrier, therefore, apparently played no role in the observed increase of the placental diffusing capacity.

In humans, birth weights are almost uniformly reported to be lower in high-altitude populations (Clegg 1978; Heath and Williams 1981); apart from hypoxia, complex genetic and nutritional factors may be involved (Cotton et al. 1980). The lower birth weights appear to be responsible for the greater placental-to-birth weight ratios observed in high-altitude populations. Whether this is associated with an increased placental diffusing capacity is not known. However, Krüger and Arias-Stella (1970) reported the number of placental cotyledons as reduced approximately by half in high-altitude populations (4.6 km in Peru) compared to sea-level controls, indicating a lesser degree of septation and probably a greater proportion of functioning parenchyma. The numbers and areas of terminal villi were also found to be greater, suggesting an increase in the surface area for gas exchange, which might be of adaptive value (Chabes et al. 1968).

Apart from possible changes in placental diffusing capacity, it must be realized that the level of maternal arterial oxygenation is critical for the O_2 supply to the fetus, and that hyperventilation is essential in raising arterial P_{O_2} under altitude hypoxia (Chap. 3). Pregnancy at high altitude has been reported to be associated with a hyperventilatory effect over and above the hyperventilation already existent on the basis of residence at high altitude (Helleger et al. 1961). More recently, studying pregnant women at an altitude of 3.1 km, Moore et al. (1982a, b) found that smaller babies were delivered by women whose minute ventilation was lower and in whom the ventilation failed to increase from early to late gestation, whereas larger babies were delivered by women whose ventilation was higher and who showed an increase in ventilation during pregnancy.

5.5 Tissue O_2 Diffusion

In spite of the various mechanisms of adaptation in the O_2 transport which have been referred to so far, the partial pressures of oxygen in systemic capillaries and the ΔP_{O_2} steps from blood to the intracellular sites of oxidative enzymes are reduced under altitude hypoxia (Fig. 2.2). Therefore, as long as oxygen uptake remains normal, there must be a mechanism that reduces the resistance that opposes O_2 flux, thus increasing the tissue conductance [G_x in Eq. (2.1)]. Theoretically, this mechanism may involve either a change in the geometry of the tissue membranes such as the ratio A/E in Eq. (2.3) increases, or an increase in the Krogh's constant of diffusion [$D_x \cdot \beta_x$ in Eq. (2.3)], or both. It will appear below that the two possibilities are supported by experimental evidence. Present knowledge, however, remains equivocal and inadequate.

5.5.1 Tissue Capillarity

An increase in the number of capillaries per unit of tissue mass (capillary density, CD; sometimes designated capillarity), thus shortening the distance through which oxygen diffuses [E in Eq. (2.3)], has long been described as a complementary mechanism of adaptation to high altitude (reviews by Stickney and van Liere 1953; Hurtado 1964; Lenfant and Sullivan 1971). Yet this remains an area of uncertainty.

The problem of capillary distribution in tissues and the effects of their spatial arrangement on oxygen diffusion remains incompletely understood and largely based upon models (Grunewald and Sowa 1977); most of these are derived from the classical Krogh's cylinder (Krogh 1922). Further, considerable variations of the capillary distribution seem to exist not only in different but also in the same tissue of one species. Under hypoxic conditions, comparative data are virtually lacking; those reported for mammals are scarce, and their interpretation is compromised by methodological differences. Besides difficulties in dealing with tissues and visualizing capillaries (see Hudlická 1982), there are problems related to the influence that growth, body size, physical activity and cold may exert on capillarization (Schmidt-Nielsen and Pennycuick 1961; Weibel 1979; Banchero 1982).

The only physiological evidence that tissue gas conductance may increase at high altitude has been provided by Tenney and Ou (1970). Using the subcutaneous gas-pocket method in rats, these investigators found that the "tissue" diffusing capacity for carbon monoxide approximately doubled after 3 weeks exposure at a simulated altitude of 5–6 km in a hypobaric chamber. Making use of certain simplifying assumptions and correcting for the effect of increased hemoglobin concentration, they deduced that a 50% increase in the capillary number should have occurred in the wall of the pocket. In this study, experiments to test the effect of acclimatization to lesser altitudes indicated no discernible effect in a 3-week period when the altitude was below 4.1 km.

Many of the morphological arguments that support an increased tissue gas conductance at high altitude are derived from studies on skeletal muscle. Because capillaries run parallel to the muscle fibers, it has been customary to describe the capillary density, CD, simply in terms of the number of capillaries per unit surface (mm^2) measured transversally to the long axis of the fibers. With such an approach, increased values of CD, suggesting more dense capillary networks and shorter diffusion distances, have been reported in various muscles of chronically hypoxic guinea pigs (Valdivia 1958), rats (Cassin et al. 1971), dogs (Banchero 1975; Eby and Banchero 1976), embryos of the bar headed goose (Snyder et al. 1981), as well as in humans (Saltin et al. 1980; biopsies from the leg muscles). It may be, however, that no new capillaries developed, but that the increased capillary density was due to a reduction in the muscle fiber diameters. With a smaller fiber diameter the capillaries are less spread apart; hence CD may increase even though the number of capillaries per muscle fiber (the capillary-to-fiber ratio, C/F) does not change. This has been found to be the case in many studies, at least when all capillaries present in the tissues were visualized with the help of appropriate histochemical reactions (for alkaline phosphatase, ATPase or periodic acid-Schiff) to show capillary endothelium. Only Valdivia (1958) found an increase in both CD and C/F, suggesting capillary proliferation at high altitude. Yet the India ink injections used in this study did not necessarily depict all capillaries similarly in the two groups of experimental and control animals. Further, the experimental group consisted of guinea pigs native to the Peruvian mountains; in these animals, greater muscular activity as well as lower environmental temperatures may have elicited capillary growth, regardless of the presence of hypoxia.

Differences in age and body size also deserve consideration. Thus, using the ATPase method in growing rats (Sillau and Banchero 1977) and guinea pigs (Sillau et al. 1980 a), the capillary density was found to decrease with increasing body mass, as a consequence of the increase in size of the fibers resulting from growth. The rates of growth and change in capillary density were not different in normoxic controls and in animals exposed for 6–14 weeks at the simulated altitude of 5.1 km (isobaric hypoxia) but kept in cages at constant, near thermoneutral ambient temperature. Hence, an increased capillary supply in skeletal muscles does not seem to be part of the normal process of adaptation of growing rats and guinea pigs to hypoxia.

An identical conclusion for cardiac muscle emerges from the data obtained in rats (Turek et al. 1972 b; Grandtner et al. 1974; Clark and Smith 1978) and guinea pigs (Rakusan et al. 1981) in either high-altitude natives, or translocated from sea level to an altitude of 3.5–4.2 km for 4–5 weeks. In the left ventricle of these hypoxic animals, no change in myocardial capillary density was noted, in comparison with normoxic controls. Only in the right ventricle was an increase in both CD and C/F found, indicating capillary proliferation and shortening of diffusion distances. Since these changes occurred only in one ventricle, they can hardly be ascribed to hypoxia per se (low P_{O_2}). More probably they were related to the heavier load on the right ventricle due to hypoxic pulmonary vasoconstriction (Chap. 4.3.2.6).

For the brain it is also questionable whether an additional supply with capillaries occurs under chronic hypoxia. Dilatation of preexisting vessels is generally evoked. This has been documented in rabbits by Opitz (1951), using ophthalmoscopy and viewing the retina as an extension of the central nervous system, and in the parietal cortex of rats by Miller and Hale (1970), using the India ink injection technique. Combining this technique with modern morphometric methods, Bär (1980) studied the microcirculatory bed of the occipital cortex of albino rats ranging in age from 14 to 120 days, and exposed for a few weeks to normobaric hypoxia simulating exposure to the altitude of 4.5 km. When applied from days 14 to 54, a period including the third postnatal week during which the processes of myelinization and capillary proliferation take place at maximal rates, hypoxia resulted in a 14% reduction in the thickness of the visual cortex, in comparison with normoxic controls. No significant cortical atrophy was observed when hypoxia was applied during later postnatal periods. In the adult rat's brain, vascular dilation occurred, but there was no evidence for the development of new capillary branches. From days 30 to 70, however, hypoxia was accompanied by the intercalation of additional endothelial cells in the existing capillaries, resulting in about 20% increase in the specific length of capillaries, hence in capillary blood volume. Functional significance and underlying mechanisms of this endothelial proliferation remain to be evaluated. At given, steady cerebral blood flow, an increase in capillary volume would result in the prolongation of mean transit time, a feature not necessarily desirable.

Vascular engorgement and tortuosity of retinal blood vessels are commonly observed in climbers at altitudes above 5 km. These signs of vascular dilatation are considered as "normal response to hypoxia"; but capillary leakage and retinal hemorrhages, as noted in one out of two subjects at 5.4 km elevation (McFadden et al. 1981), are "abnormal reactions", and are classified as high-altitude retinopathy. Underlying mechanisms are disputed, but the high blood flow and increased intravascular pressure under physical exertion with possible hypoxic damage to the vascular endothelium may play important roles. For further details, see Heath and Williams (1981).

5.5.2 Myoglobin

It is well known that myoglobin (Mb) is a respiratory pigment the affinity of which for oxygen lies intermediate between blood hemoglobin and tissue cytochrome oxidase. Essentially found in the red fibers of skeletal muscles and in cardiac muscle, Mb is believed to have a dual function in the O_2 supply (Kreuzer 1970; Wittenberg 1970; Wittenberg and Wittenberg 1981). First, by the very fact of its reversible combination with oxygen, Mb acts as a short-term oxygen store, which can be used in situ, virtually with no time lag. In steady state, this storage function makes no contribution to oxygen supply. Second, being closely packed intracellularly, and by constantly taking up and giving off oxygen from and to each other, functional Mb molecules are considered to speed the inward diffusion of oxygen across the cytoplasm. This physicochemical phenomenon, called Mb-facilitated O_2 diffusion, can be viewed as increasing the effective value of the Krogh's diffusion constant [$D_x \cdot \beta_x$ in Eq. (2.3)]. Conceivably, therefore, the increased Mb concentration at high altitude which has long been described (e.g., Stickney and van Liere 1953; Hurtado 1964) might be of adaptive significance. Yet this also remains an area of uncertainty.

That Mb enhances the flux of oxygen across myoglobin-containing liquid phases is documentated in vitro (see the reviews quoted above), though underlying mechanisms are not clarified (Cole et al. 1982). The in vivo effect of myoglobin under truly physiological circumstances is as yet unknown. Recent evidence, however, indicates that Mb may be important in maintaining muscle function under conditions of hypoxia. Thus, in anesthetized dogs made to breathe a hypoxic gas mixture (10% O_2 for 20 min), Cole (1983) found that both the oxygen uptake and tension generation of the electrically stimulated in situ gastrocnemius-plantaris decreased (by about 40%) when Mb was converted to nonfunctional forms, incapable of reversible combination with oxygen. This occurred after administration of hydrogen peroxide, a treatment which did not affect blood flow, perfusion pressure or mitochondrial oxidative phosphorylations. This observation in vivo is in keeping with the in vitro demonstration that the conversion of intracellular Mb to irreversible high oxidation states results in the reduction of oxygen uptake when oxygen supply is limiting (Wittenberg et al. 1975).

That Mb concentration "increases somewhat in the muscles of man and of domestic animals kept at high altitude and is greater in wild animals living at high altitude than in their lowland relatives" has been reviewed by Wittenberg (1970). Reports, however, are scanty, and concerned with a limited number of muscles in a few mammalian species. Further, the data are controversial, probably because of the significant differences that age, nutritional status, exercise, and cold (Holloszy 1975), as well as the nature and duration of hypoxia, produce on Mb concentration (Vaughan and Pace 1956; Anthony et al. 1959).

In cardiac muscle, the Mb concentration was found to be increased by 15–50% in the hearts of rats either acclimatized to elevations of 3.8–5.5 km (Vaughan and Pace 1956; Anthony et al. 1959), or born and raised at 3.8 km (Vaughan and Pace 1956). These earlier reports, however, made no distinction between right and left ventricles. It appears from later studies in which the distinction was made that hypoxia per se does not affect the Mb concentration in cardiac muscle. In growing rats (Turek et al. 1973 b) and guinea pigs (Bui and Banchero 1980), no change was found in the Mb concentration of the left ventricle in the course of a 4–14 weeks exposure at the simulated altitudes of 3.5–5.1 km. In these hypoxic animals, as compared with controls, a higher Mb concentration was found only in the right ventricle, together with hypertrophy. Hence, not the low P_{O_2} per se but the heavier work load on the right ventricle induced by the hypoxic constriction of the pulmonary vascular bed (Chap. 4.3.2.6) was most probably the stimulus.

Things may be not that simple, however. Thus, comparing mountain pocket gophers (3.15 km) with valley relatives (0.25 km) Lechner (1976) found a nonsignificant 7% increase in the Mb concentration of the two ventricles excised from the high-altitude native animals. Body mass was similar in the two groups, but age could not be determined when this information might have been of value. Indeed, in growing rats, Turek et al. (1973 b) noted that the myocardial Mb concentration was better correlated with the age than with the body mass. On the other hand, these authors observed that a cardiomegaly, most pronounced in the right ventricle, developed in the rats born and raised at 3.5 km altitude; yet in comparison with sea-level controls, the Mb concentration was unchanged in the right ventricle, but decreased in the left ventricle. Left ventricular "hypotrophy" with no change in Mb concentration was found in the hypoxic guinea pigs studied by Bui and Banchero (1980). Underlying mechanisms and significance of these changes are obscure.

Turning to skeletal muscles, much of the available evidence indicates that the Mb concentration increases at high altitude. This has been shown for various muscles in humans (Reynafarje 1962), dogs (Hurtado et al. 1937; Gimenez et al. 1977), rats (Vaughan and Pace 1956; Anthony et al. 1959), pocket gophers (Lechner 1976), and guinea pigs (Tappan and Reynafarje 1957) either native or acclimatized to altitudes ranging from 3.1 to 5.5 km, in natural conditions or in a hypobaric chamber. The opposite observation that Mb concentration decreases when a high-altitude animal is brought down to sea level has been reported for the gracilis muscle of alpacas (Reynafarje et al. 1975).

What remains uncertain is whether hypoxia per se, or exercise, or cold, or another factor yet undetermined was responsible, at high altitude, for the observed increase of the Mb concentration in skeletal muscle. Thus, on the basis of associated changes in body weight and muscle weights, Anthony et al. (1959) believed that the increment in Mb concentration that they observed may have been more apparent than real, and possibly attributable to dehydration.

The nature and duration of the hypoxic exposure, as well as the type of muscle, and the age or body size, appear to be crucial. Thus, no difference between control and experimental rats was found when discontinuous hypobaric hypoxia (Bowen and Eads 1949; Poel 1949) or continuous isobaric hypoxia (Sillau and Banchero 1977) were used to simulate altitudes ranging from 5.1 to 7.6 km. In growing guinea pigs exposed to isobaric hypoxia corresponding to the altitude of 5.1 km as well as in growing controls, Sillau et al. (1980a) noted that the Mb concentration in gastrocnemius and soleus muscles increased linearly as a function of body mass; a clear-cut greater increase in the hypoxic group was significant only in the soleus muscle after 14 weeks of exposure. Three months of exposure at 0.5 atm were necessary for the changes in Mb concentration to occur in a muscle of the hind legs in the rats studied by Costa and Taquini (1970).

Apart from the changes in the concentration of Mb, one may wonder whether there are changes in the reaction rates of O_2 uptake and release by the myoglobin molecule during altitude hypoxia. In rats acclimatized at the simulated altitude of 6 km for 4–12 weeks, and in control animals kept at sea level, Strother et al. (1959) compared the rate constants for purified extracts of myoglobin from cardiac muscle and various skeletal muscles. They observed no difference for skeletal muscle extracts of acclimatized versus control animals. For heart Mb extracts, the results suggest that the myoglobin from acclimatized animals might release more of its stored oxygen at a given low oxygen pressure.

5.6 Summary

This chapter examines whether the structural characteristics and functional properties of the gas-exchanging surfaces (barriers) in the body improve in the course of altitude hypoxia in order to maintain oxygen consumption with declining oxygen tension.

Taking the oxygen transport across the avian eggshell as a purely diffusive process in a gas phase, it becomes clear that the ambient partial pressure of oxygen (P_{O_2}) and atmospheric pressure (P_B) exert independent effects on gas exchange of the avian embryo. With a fixed pore geometry, the effective conductance of the eggshell is increased at high altitude because the diffusion coefficient of gases is inversely proportional to P_B. This offers the potential advantage of compensating for the drop of the ambient P_{O_2}, but brings with it the threat of hypocapnia, increased loss of water vapor, and therefore dehydration. For birds breeding in mountain environments, it seems that the regulation of loss of water vapor by a reduction in the pore area, hence by a reduction of eggshell conductance, has been more important than maximizing O_2 transport to the embryo.

Taking the alveolar-arterial P_{O_2} difference as an overall appraisal of the O_2 transport across the lung, a common finding is that this difference is lower, both at rest and during exercise, in native highlanders than in sea-level natives, both studied in their native environments. Also reported in some other mammals, this phenomenon means that the $\dot{M}_{O_2}$-specific conductance across the lung is increased at high altitude. Independent or simultaneous changes in the following three components of the alveolar-arterial P_{O_2} difference may be involved: (1) venous blood admixture, (2) ventilation-perfusion inequality, and (3) diffusion limitation. Little is known to date, mainly due to technical difficulties.

Not only the contribution of the venous blood admixture (at steady magnitude) to the widening of the alveolar-arterial P_{O_2} difference decrease under altitude hypoxia, due to the very shape of the blood O_2 equilibrium curve, but also its magnitude may be reduced. Opposite findings have been reported concerning the hypoxic changes in ventilation-perfusion inequality. Diffusion limitation appears to be reduced both in sojourners and natives at high altitude. In the former, however, the reported increase in lung-diffusing capacity can be explained by the increased hemoglobin concentration. Only in native highlanders was the diffusing capacity of the pulmonary membrane itself found to be raised above sea-level values. Underlying mechanisms are not clear. Experiments in mammals, fish, and amphibians, indicate that (1) in young animals, the growth of the lung with respect to body growth can be accelerated toward normal adult dimensions, whereas (2) adult animals do not seem to be able to adapt their lung structure to hypoxia.

Limited data on pregnant ewes and pregnant guinea pigs suggest that adaptive changes occur under altitude hypoxia in the complex O_2 transfer from maternal to fetal blood. In particular, the placental diffusing capacity may increase. Both an increase in the surface area for gas exchange and a reduction in the diffusion distance between maternal and fetal blood have been advocated, but not fully documented.

For a long time it has been thought that additional tissular capillary supply occurs under chronic hypoxia, thereby shortening the distance through which oxygen diffuses, hence improving the O_2 conductance from capillary blood to cells. Recent studies make it necessary to reconsider this view. Taking into account the interacting effects of cold exposure, exercise load, and changes in age and body size, it appears that hypoxia per se has no specific effect on tissue capillarity. Increased capillary supply does not seem to be part of the normal process of adaptation to hypoxia.

It also remains uncertain whether hypoxia per se, or work load, or cold, or some other factor yet undetermined, is responsible for the increase of myoglobin concentration that may occur in some skeletal muscles in the course of exposure to altitude hypoxia.

Chapter 6

Biochemical Changes*

The final aspect of the respiratory processes under altitude hypoxia is the intracellular biochemical changes that may be adaptive in nature. Paul Bert was the first to postulate in *La Pression Barométrique* that body tissues may gradually alter their cellular metabolism during acclimatization to high altitude: "It may very well be that mountain dwellers have a better-regulated (cellular) machine, which is considerably more efficient, and consequently, for the same dynamic expenditure, requires a smaller absorption of oxygen and of food" (Bert 1978, p 1003). Since that time, conflicting descriptive data have accumulated on the problem of cellular and metabolic adaptation to hypoxia. Development of a suitable in vivo model for the study of the metabolic adjustments to altitude hypoxia has proved to be difficult. The animal species and age, the type (continuous versus discontinuous), the degree (moderate versus severe), the nature (hypocapnic versus normo-, or hypercapnic) and the duration (short versus prolonged or chronic) of the hypoxic exposure, the possible interaction between hypoxia and cold or exercise training, the nutritional state (feeding versus fasting) at the time of measurements, even the methods of nursing the animals, sampling and manipulating the tissues, all these factors can be invoked to explain the differences in results. Also, it may be that, in the biochemical approach of the adaptation to hypoxia, "... failure to understand how a steady-state flux through a (metabolic) pathway can be maintained and how it can be changed from one rate to another ... has led and is still leading to a considerable wastage of effort in experiments on metabolic control" (Newsholme and Crabtree 1981).

This chapter attempts to summarize present knowledge. So far, however, the cellular adjustments to altitude hypoxia are poorly described and not understood. Some aspects of the problem, and historical background can be found in several reviews (e.g., Stickney and van Liere 1953; Knox et al. 1956; Barbashova 1964; Cohen 1973; Timiras 1977; Wegener 1981) and books (Hochachka and Somero 1973; Hochachka 1980; Heath and Williams 1981).

6.1 Oxygen Utilization in Cells

In practical terms, the physiologist studying an organism in the steady state, and measuring oxygen uptake, considers he is getting a good estimate of the oxygen consumed at the level of mitochondrial respiratory chain, where useful energy is liberated in the form of adenosine triphosphate through phosphorylating oxidations driven by electron-transfer oxidase. To a large extent this is true. In the or-

* Written in collaboration with C. Leray

ganism, however, there are processes different from those of mitochondrial respiration, which utilize molecular oxygen (cyanide-insensitive oxygen uptake) but are not directly involved in the production of energy. In various subcellular organelles (de Duve 1966, 1983) they function in the biosynthesis of a number of biologically important molecules, by direct insertion of one or two oxygen atoms into the products through the action of specific enzymes (oxygenases and hydroxylases). Accordingly, the biochemical consequences of hypoxia on phosphorylating oxidations and nonphosphorylating oxidations will be examined, respectively under the following subheadings: Bioenergetic Adaptations and Nonbioenergetic Adaptations.

6.2 Bioenergetic Adaptations

6.2.1 Introductory Remarks

It is well established that adenosine triphosphate (ATP) is the most important source of energy for most biological energy-requiring processes. Hydrolytic removal of one phosphate group from ATP yields adenosine diphosphate (ADP) and releases 10 to 12 kcal $\cdot$ mol^{-1}; splitting off a second phosphate group from ADP yields adenosine monophosphate (AMP) and makes further energy available. In steady, normal functioning of the cell, ATP production is equal to ATP utilization and the following ratio

$$AEC = \frac{[ATP]+0.5[ADP]}{[ATP]+[ADP]+[AMP]} \tag{6.1}$$

is about 0.8–0.9 depending on the tissue under consideration, with AEC being the adenylate energy charge. The regeneration of ATP depends on the catabolism of carbohydrates, fats, and proteins. This is schematically reminded in Fig. 6.1,

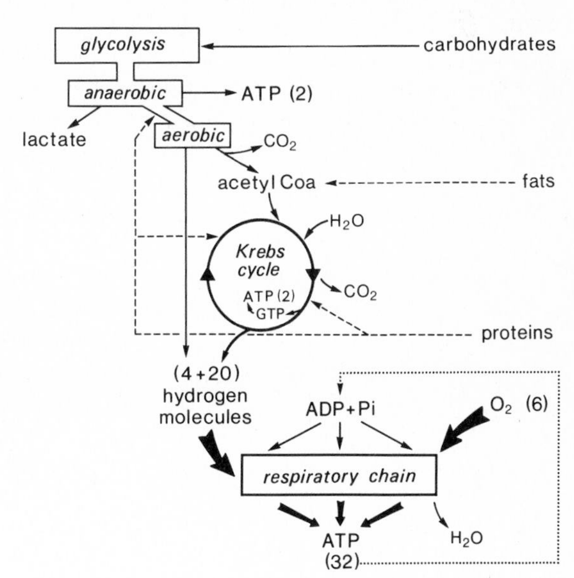

Fig. 6.1. Oversimplified representation of the metabolic transformations involved in the production of ATP from nutrients

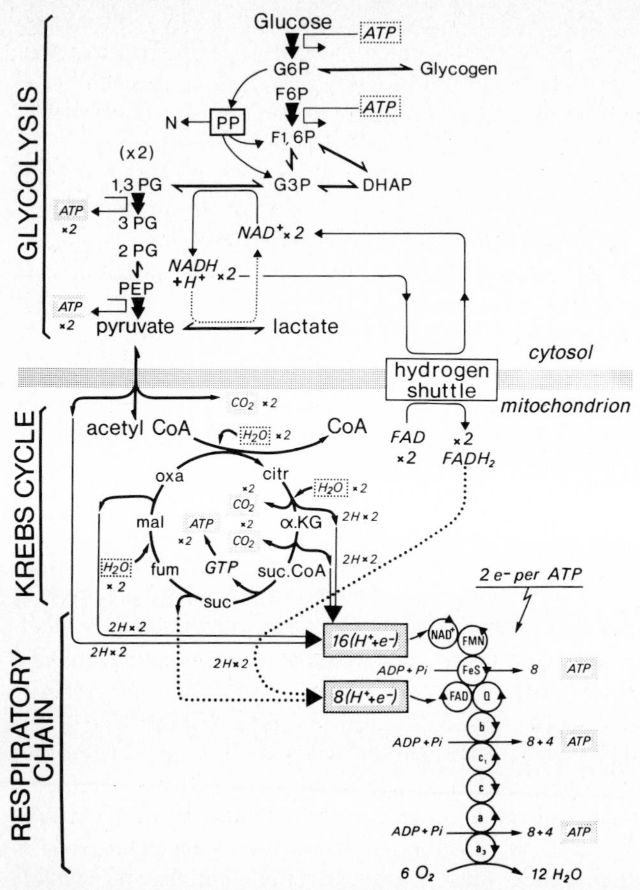

Fig. 6.2. The three stages of oxidative metabollsm: glycolysis, Krebs cycle, and oxidative phosphorylations in the respiratory chain. Note the sources of hydrogen molecules driving the respiratory chain, the various steps of ATP (adenosine triphosphate) generation in the cytosol and mitochondrion, and the mechanisms involved in the maintenance of redox status

which shows that the Krebs cycle is the final common pathway into which all the fuel molecules are ultimately degraded. Figure 6.2 demonstrates the catabolism of carbohydrates.

Focusing on the respiratory chain (Figs. 6.1 and 6.2), the whole machinery of cellular respiration may be viewed as a black box with three open entries for *oxygen* (by diffusion from outside), *fuel* in the form of hydrogen molecules (i.e., protons + electrons, or "reducing equivalents," which are derived from substrates, and transferred through the action of coenzymes, like NAD^+, the nicotinamide adenine dinucleotide, also named diphosphopyridine nucleotide), and *energy acceptors* (ADP and inorganic phosphate, Pi), and two output channels for *water,* and for the *energetical product,* ATP. Note that (1) oxygen is the terminal electron acceptor, at the end of respiratory chain, (2) there is a greater energy yield when oxygen is available than under anaerobic condition (36 ATP mol-

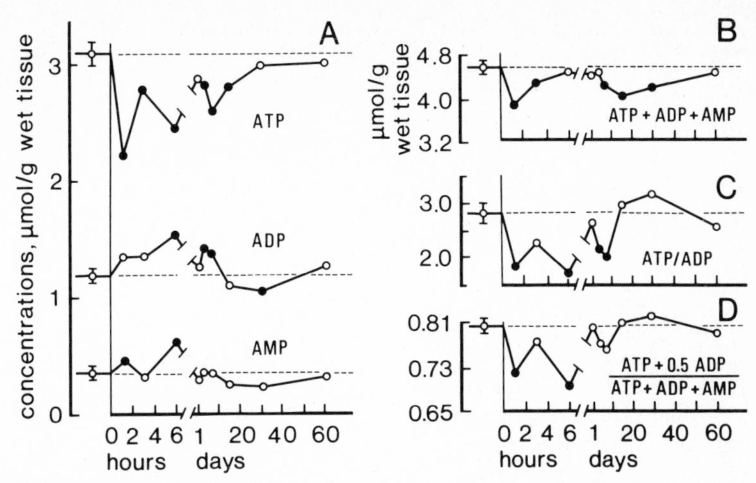

Fig. 6.3. Changes in the energy status of rat liver in the course of exposure to the altitude of 3.8 km. (After Reed and Pace 1980). *Points* represent mean values (± S.E.; n = 12). *Filled symbols* indicate a significant difference from sea-level values shown on the left of time zero. ATP, ADP, and AMP: adenosine tri-, di-, and monophosphate

ecules against 2), (3) there is no problem of *redox balance*[7] when oxygen is available (reduced coenzymes are reoxidized at the level of the respiratory chain) whereas, under anaerobic conditions, the reduced coenzyme NADH that originates from glycolysis must be reoxidized to NAD^+ via a purely cytosolic process, through the action of lactate dehydrogenase; the lactate, which then accumulates at the detriment of pyruvate, and yields acidosis, must be withdrawn into the circulating blood.

6.2.2 A Typical Experiment

In a recent study (Reed and Pace 1980), young adult male rats of the Long-Evans strains were either depressurized for 1–6 h in a hypobaric chamber operated at 475 Torr, or transferred for 1–60 days in the White Mountain research station at 3.8 km (barometric pressure ranging from 470 to 490 Torr). Enzymatic analyses were performed on extracts from liver quenched in liquid nitrogen. Some of the results are shown in Figs. 6.3 and 6.4.

Judging from the changes in adenosine phosphates (Fig. 6.3), a depression of cellular energy status occurred with short-term hypoxia. Both ATP/ADP and adenylate charge ratios (panels C and D, respectively) indicate that by 1–15 days of high-altitude exposure, there has been a return to near sea-level values. The major

7 Each oxidative step in metabolism must be balanced by a reduction step. Estimates of redox changes are based on determinations of the $NADH/NAD^+$ ratio, either through spectrophotometric or fluorimetric measurements of spectral characteristics of the dinucleotide (which are altered by its redox state), or through measurements of substrates in equilibrium with, and coupled to the nucleotide reaction (for details and references see Williamson et al. 1967; Cohen 1972; Hochachka 1980)

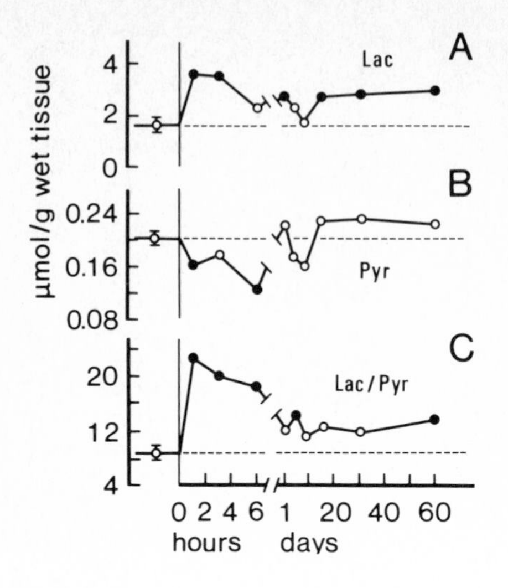

Fig. 6.4. Changes in lactate (*A*) and pyruvate (*B*) concentrations, and in the lactate/pyruvate ratio (*C*) in rat liver cells in the course of exposure to the altitude of 3.8 km. (After Reed and Pace 1980). *Points* represent mean values (±S.E.; n = 12). *Filled symbols* indicate a significant difference from sea-level values shown on the left of time zero

role of lactate accumulation as a process to regenerate NAD^+ from NADH is evidenced by its consistent elevation throughout the altitude exposure period (Fig. 6.4 A). Aside from providing pragmatic information on substrate shifts per se, the increased lactate/pyruvate ratio observed during the first hours of hypoxic exposure (Fig. 6.4 C) is an indication of an early increase in reduction of the extramitochondrial $NADH/NAD^+$ system; this was verified by direct fluorimetric measurements. On the other hand, the authors found a significant decrease in the total adenosine phosphates (Fig. 6.3 B) and pyridine nucleotides. They discussed the point that, besides hypoxia, other factors including the stress of transportation and hypophagia, and possible alterations in other metabolic pathways, may have been responsible for some of the observed changes. In their opinion, however, the fact "that the various criteria of cellular energy status and redox status returned to near-normal values indicates that, under conditions of the current experiments, intracellular compensation, and acclimatization did occur." Yet, as will be seen, many uncertainties remain about underlying mechanisms.

6.2.3 Respiratory Chain

The electron transport is the essential link between O_2 consumption and ATP production. It consists of a cascade of oxidation-reduction couples by which the free energy difference between the electrons derived from foodstuffs and terminal oxygen is degraded in a stepwise fashion. These are macromolecular assemblies, which are tightly associated with the inner mitochondrial membrane, in a dense and regular array (Hinkle and McCarthy 1978; Racker 1980). The inner mitochondrial membrane may be viewed as a mosaic of these functional units. Their dense arrangement suggests that the number of such units is proportional to the surface area of the inner membrane, including its infoldings, the mitochondrial cristae (Weibel 1979). Thus, not only the enzymic properties of the components of the respiratory chain, but also the number and size of mitochondria are potential adaptates to hypoxia.

Table 6.1. Metabolic states and associated steady oxidation-reduction percentages of some components of the respiratory chain. (After Chance 1957; rat liver mitochondria in vitro at 25 °C; substrate = hydroxybutyrate; phosphate acceptor = ADP)

State	Starved	Active	Resting	Anaerobic
Concentrations of				
Substrate	0	High	High	High
ADP	High	High	Low	High
Oxygen	+	+	+	0
ATP	Low	Low	High	Low
Respiratory rate	0	Fast	Slow	0
Rate-limiting factor	Substrate	Respiratory chain	ADP	Oxygen
% Steady reduction				
of NADH	0	53	>99	100
Flavoprotein	0	20	40	100
cyt b	0	16	35	100
cyt c	0	16	14	100
cyt a	0	< 4	0	100

6.2.3.1 Critical Oxygen Tension

Current views on the O_2 dependence of oxidative phosphorylations have been exposed in many reviews and articles (e.g., Chance 1957, 1965, 1977; Jöbsis 1964, 1972, 1979; Wilson et al. 1977a, b, 1979a, b). However, the concept of the critical value of P_{O_2} required for normal mitochondrial metabolism remains open. First, the activity of the respiratory chain is affected by many factors. Table 6.1 shows the prime importance of the substrate (i.e., protons + electrons) and ADP concentrations, apart from that of oxygen: at given O_2 availability (tension), oxygen may very well be present in excess at rest or in starved condition, whereas it may become limiting in active condition. Thus, a so-called "critical P_{O_2}" will not reflect the intrinsic characteristics of mitochondrial respiration itself, but rather those of the energy metabolism of the tissue (Sugano et al. 1974). Second, due to diffusion barriers, quantitative assessment of the critical O_2 tension is a matter of site of measurement.

a) In mitochondrial particules, isolated in vitro from varied tissues, the rate of oxygen consumption and the level of cytochrome oxidase reduction are known to remain constant until oxygen tension is extremely low (ca. 2 Torr; Chance 1957). At this critical P_{O_2}, respiration begins to decrease, a half-normal oxygen consumption being reached at a P_{O_2} of about 0.5 Torr in the assay cuvette.

b) In cell suspensions, the rate of O_2 diffusion may be a critical factor for intracellular O_2 supply to mitochondria during hypoxia. Thus, the level of cytochrome oxidase reduction was found to increase progressively in isolated hepatocytes when the value of P_{O_2} in the bathing medium became less than 15 Torr (Jones and Kennedy 1982). This observation indicates that a major limiting factor in O_2 supply to support mitochondrial function may be the need for a capillary P_{O_2} high enough to maintain saturation of cytochrome oxidase in the presence of intracellular P_{O_2} gradients.

Similar results have been obtained by Wilson et al. (1977a, b, 1979a) in cultured cells from various organs, as well as in tumor cells or protozoans. These authors found that O_2 uptake remained remarkably stable until the value of P_{O_2} was less than 12–15 Torr. However, as P_{O_2} declined from its normoxic value, cytochrome c became more and more reduced and ATP/ADP ratio decreased. Through the modification of cytosolic [ATP]/[ADP][Pi] and intramitochondrial $NADH/NAD^+$ ratios, mitochondrial phosphorylations might be capable of not only sensing cellular P_{O_2} throughout the physiological range, but also of transmitting this observation to other metabolic pathways and to circulatory convective mechanisms (Wilson et al. 1979b, 1980).

c) In perfused tissues, Stainsby and Otis (1964) found that muscle O_2 uptake began to decrease when the venous P_{O_2} was about 25 Torr in resting condition, but 10 Torr for contracting muscles. Most certainly the number of open capillaries increased in contracting muscles, thereby decreasing the thickness of the diffusion barrier.

Unlike other intact tissues in which intracellular P_{O_2} depends on O_2 delivery via vascular perfusion, and on the presence of a significant diffusion barrier, alveolar cells of isolated lungs artificially ventilated and perfused may constitute a suitable model for the study of metabolism. Thus, Fisher and Dodia (1981) reported that a value of alveolar P_{O_2} less than 1 Torr is required to produce significant change in the ATP concentration and redox balance in the lung parenchyma.

d) In conscious animals, there is evidence that the electron transfer system in mitochondria is sensitive to changes of P_{O_2} within the normoxic range. Thus, using a noninvasive reflectance spectrophotometric technique, Jöbsis studied the level of reduction of cytochrome oxidase in the intact cerebral cortex of either unanesthetized rabbits with chronically implanted cranial windows (Rosenthal et al. 1976) or decerebrated cats (Jöbsis et al. 1977). The reduction level of cytochrome oxidase was found to be 20% in normoxic animals, to increase under hypoxia, and to decrease when the animals were given some oxygen to breathe. A scarcity of hydrogen ions under normal function has been advocated to contribute to the high reduction level of the terminal component of the respiratory chain (Jöbsis 1979).

At variance with Jöbsis's observations, the optical measurements of oxygen delivery and consumption in gerbil cerebral cortex, reported by Bashford et al. (1982), show that the components of the respiratory chain in vivo do not become reduced until the fraction of oxygen in the inspired gas falls below 4%, when virtually all of the cerebral hemoglobin is deoxygenated. This would indicate that the properties of mitochondria and cytochrome oxidase in cerebral cortex of gerbils may resemble those found in isolated mitochondria. Reasons for such species differences are unclear.

6.2.3.2 Hypoxic Changes

Since the respiratory chain is directly concerned with O_2 utilization, it is appropriate to examine whether some regulatory responses occur in the course of hypoxic exposure. To date, however, the collected data remain scanty, phenomenological, and conflicting.

a) Oxidative capacity: Respiratory rates in active state and in u states[8] have been reported either to increase twofold (Mela et al. 1976) crease by 10–30% (Strickland et al. 1962; Nelson et al. 1967; Gold et Costa et al. 1982) in mitochondria of heart and liver isolated from rats ma oxic. The experimental conditions, however, were different in the two gr study. In the first experiment (Mela et al. 1976), the animals breathed a h gas mixture for 30–90 min at sea level; the equivalent altitude was close to 8–10 km and P_{CO_2} was prevented from falling in the hyperventilating animals by addition of 5% CO_2 to the inspired gas. Hence, this study was concerned with the acute effects of a severe normocapnic hypoxia. In the second group of experiments, the animals were studied in a hypobaric chamber operated at simulated altitudes of 4.4–7.6 km for periods of time ranging from 3 days to 11 weeks, thus being exposed to prolonged hypocapnic hypoxia. Hypophagia and loss of body weight were consistently observed. Further, identical decreases in oxydative capacity occurred in pair-fed rats maintained near sea level and supplied with the same average daily quantity of food that the altitude rats voluntarily consumed. Consequently, the observed decreases may be related to a state of semi-starvation (Gold et al. 1973). For the animals living naturally at high altitude, the scanty available data are also contradictory. Ou and Tenney (1970) found an increased oxygen uptake rate in mitochondria isolated from heart tissue of cattle born and raised at 4.25 km, in comparison with sea-level cattle of the same breed. No significant difference was observed by Reynafarje (1971) on comparing respiratory rates of either resting or active mitochondria isolated from heart or liver of two groups of guinea pigs, one native to 4.5 km and the other to sea level.

b) Phosphorylating capacity: The above studies all agree in reporting no significant change in the amount of O_2 required to convert a given amount of ADP to ATP. On the other hand, Reynafarje (1971) found that the apparent Michaelis-Menten constant for ADP was significantly smaller in the cardiac and hepatic mitochondria isolated from guinea pigs native to high-altitude than to sea-level. This would suggest that, in the presence of a given amount of ADP, mitochondria from high-altitude animals were able to generate more energy in the form of ATP.

c) Enzyme activities: In regard to the activity, or concentration, of the individual components of the respiratory chain during hypoxia, some studies have showed no change, others an increase, and others a decrease. Most data are concerned with two of these components, succinate dehydrogenase and cytochrome oxidase in various tissues. To make the values from different laboratories comparable, these data are expressed in Fig. 6.5 as the dimensionless ratio of the values in hypoxia over those in control normoxia (H/N).

As for succinate dehydrogenase (SDH; left part of Fig. 6.5), an increased activity at high altitude is reported in almost all the studies. In contrast, results are conflicting in regard to cytochrome oxidase (cyt aa_3; right part of Fig. 6.5).

---

8 Oxidative phosphorylation becomes *uncoupled* from electron transport whenever anything happens (as the use of dinitrophenol, for instance) that makes the inner mitochondrial membrane permeable to protons. The production of ATP then ceases while both electron transport and the reduction of O_2 to H_2O increase

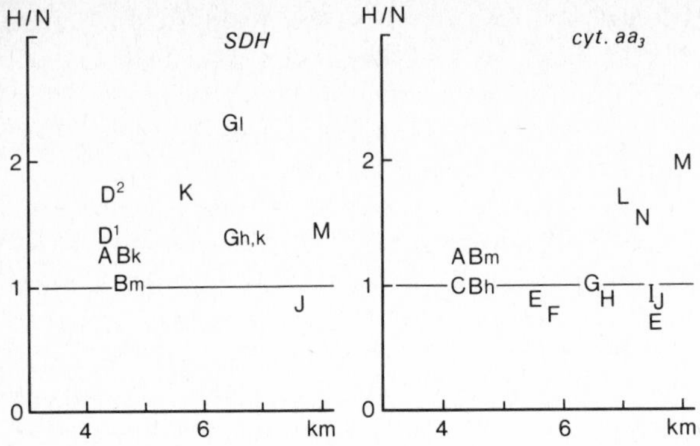

Fig. 6.5. Effects of hypoxia on concentrations (*A, E, F, H, L*) or activities (*other capitals*) of succinate-dehydrogenase (*SDH*) and cytochrome oxidase (*cyt* aa_3) in mitochondria from various mammalian organs. *Ordinates* relative changes expressed as the ratio of hypoxic (*H*) over control normoxic (*N*) values; *abscissa* altitudes. Abbreviations in following list: *h.a.* = high altitude (in mountains); *h.h.* = hypobaric hypoxia in a chamber; *n.h.* = normobaric hypoxia (by changing inspired F_{O_2} near sea level). *A* – Ou and Tenney (1970), heart of cattle natives to h.a. *B* – Tappan et al. (1957), heart (*h*), kidney (*k*), and muscle (*m*) of guinea pigs natives to h.a. *C* – Camba et al. (1980), liver of guinea pigs natives to h.a. *D* – Harris et al. (1970), heart of rabbits (*1*), guinea pigs and dogs (*2*) natives to h.a. *E* – Kinnula (1976), heart and liver of rats exposed to n.h. for 1–2 d. *F* – Kinnula and Hassinen (1977), heart and liver of rats exposed to n.h. for 1–2 d. *G* – Shertzer and Cascarano (1972), heart (*h*), liver (*l*), and kidney (*k*; cyt aa_3 values are similar in all three organs) of rats under h.h. at 0.5 atm for 14 d. followed by 14 d. at 0.4 atm. *H* – Albaum and Chinn (1953) brain of rats exposed to h.h. for 1–3 months. *I* – Gold et al. (1973), liver and kidney of rats exposed to h.h. for 7 d. *J* – Gold and Costello (1974), hearts of rats exposed to h.h. for 7 d. *K* – Susheela and Ramasarma (1973), liver of rats exposed to h.h. for 4 h. *L* – Kinnula and Hassinen (1981), intestinal epithelial cells of rats exposed to h.h. for 2 d. *M* – Hazen and Kuznets (cited by Meerson 1975, p 100), cerebral cortex of rats exposed to intermittent h.h., 6–8 times for 2 h. *N* – Smialek and Hamberger (1970), brain of rabbits exposed to n.h. for 6 h

When comparing animals native to high altitude with sea-level controls (data points A, B, and C), either no change or a slight increase in the activity of cytochrome oxidase has been reported. It is worth noting that Camba et al. (1980, point C) found no increased activity when comparing two groups of guinea pigs at normal ambient temperature, one native to high altitude in Peru (4.25 km) and the other from near sea level (0.15 km); in the two groups, however, the cytochrome oxidase activity increased under cold exposure.

A decrease in the cytochrome oxidase activity has been reported in almost all the studies on rats from sea level exposed to simulated altitudes above 5 km (data points E to J in Fig. 6.5, right part). The function served by such a decreased activity, also observed in a number of eukaryotic cells made hypoxic in vitro, is not clear (Robin 1980). In fact, not only hypoxia, but a state of semi-starvation may have been implicated (Gold et al. 1973, point I in Fig. 6.5; Kinnula and Hassinen 1977, point F).

Finally, three data points (L, M, and N) indicate an increased cytochrome oxidase activity under very severe hypoxia. Points M and N will be considered later. Point L refers to the work of Kinnula and Hassinen (1981) on the villous epithelium of the small intestine (a tissue with high cellular turnover) of pair-fed rats.

The half-life time of the intestinal cells was found to increase within 2 days, in the hypoxic group than in fasting normoxic controls. The observed increa the activity of cytochrome oxidase activity may have been related to the lo life of the cells, which allowed the synthesis of larger amounts of mitochonc components.

d) Mitochondrial protein synthesis: Its rate has been reported to increase within 10 days in the heart and brain of rats exposed to an altitude of 5–7 km in a hypobaric chamber for 5–6 h daily (Merson 1975). Activation of mitochondrial DNA and RNA synthesis was concomitantly observed, thus suggesting activation of the genetic apparatus of the cells. Similarly, the study of Smialek and Hamberger (1970) showed that hypoxic increase in the activity of cytochrome oxidase (point N in Fig. 6.5) was combined with an activation of the mitochondrial protein synthesis. In so far as brain RNA and protein synthesis may increase under the stimulation of neurons of the central nervous system (Meerson 1975), one may wonder whether a hypoxic "cortical arousal" (known to result from the stimulation of the reticular activating system by hypoxia via the arterial chemoreceptors; Hugelin et al. 1959), was involved in the responses shown by points M and N in Fig. 6.5 (right part).

e) Number and size of mitochondria: As already mentioned, the inner membrane and cristae of a mitochondrion can be viewed as a mosaic of respiratory functional units, the activity of which is thought to be proportional to the surface area of the membrane and cristae. Based on this premise, morphological techniques have been used in an attempt to assess the possible adjustments of the respiratory chain to hypoxia. Thus, Ou and Tenney (1970) compared the hearts of two groups of domestic cattle, one native to high altitude (4.25 km) and the other from sea level. They found a 40% increase in the number of mitochondria in the high-altitude hearts. The size of mitochondria, on the other hand, was the same as at sea level. In contrast, Kearney (1973) was unable to show any quantitative difference between the myocardial mitochondria of two groups of rodents (rabbits and guinea pigs), one native from 4.3 km and the other from sea level. Hence, results so far available are not conclusive.

The "optimal" distribution of mitochondria in cells, between the sites where ATP is being utilized and the entrance ports for O_2 and metabolites (Weibel 1979) appears to be changed in an apparently functionally meaningful way when young dipterans are exposed to low O_2 pressure (Thyberg et al. 1982). Whether such a phenomenon occurs in vertebrates exposed to altitude hypoxia remains to be investigated.

6.2.4 Krebs Cycle

If the respiratory chain is to yield any energy at all, then it must be supplied with hydrogen molecules. These come essentially from the Krebs cycle (Figs. 6.1 and 6.2).

Alterations in the Krebs cycle metabolism and its related enzymes during altitude hypoxia have been studied minimally. In a recent study, Hochachka et al. (1983) found that the activities of citrate synthase (CS) and hydroxyacylCoA dehydrogenase (HOAD) of the heart (left ventricle) were elevated in three high-al-

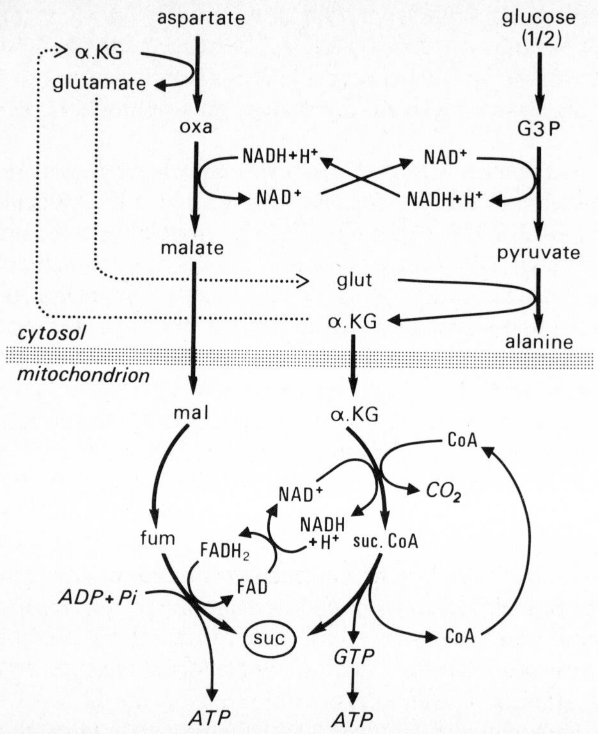

Fig. 6.6. Schematic representation of the metabolic pathways involved in the anaerobic accumulation of alanine and succinate, supplementary ATP generation and maintenance of the redox balance in both the cytosol and mitochondrion. (After Hochachka et al. 1975). See text for explanation

titude mammalian species compared with other mammals from low altitude. Oxidative capacity of the heart, as indexed by CS activity $\cdot$ g^{-1}, was by far the highest in taruca (a wild deer, which ranges up to 6.0 km altitude), but it also scaled upward in domestic camelids (llama and alpaca) compared with sea-level ox. On the other hand, a fat-based metabolic organization was found strikingly active in the camelids, occurring at about two fold higher activities than in taruca heart, 7–10 times higher than in ox.

Earliest studies were phenomenological and conflicting. Berry et al. (1957) found no change, but Mensen de Silva and Cazorla (1973) reported an increase of the citrate concentration in the heart of guinea pigs native to high altitude (4.3 km in the Andes) compared with sea-level animals. The citrate concentration was also found to increase slightly in rats translocated for 85 days from low altitude to 3.5 km in the Alps (Vergnes 1973).

Mensen de Silva and Cazorla (1973) found that the succinate concentration in various tissues was elevated in high-altitude guinea pigs, compared with sea-level animals. This observation can be correlated with the hypothesis that some amino acids may function as anaerobic fuel for cells (Hochachka et al. 1975). The process, schematically shown in Fig. 6.6, requires (1) the transamination of aspartate with α-KG to form glutamate and oxaloacetate, (2) the concomitant oxidation of glucose to pyruvate, with accompanying transamination of pyruvate with glutamate to form alanine and α-KG, (3) the passage of α-KG and malate in the mitochondrion, where their metabolism generates ATP. The redox status of both the cytosol and mitochondrion remains unaltered, while two ATP molecules per pyruvate utilized (four per starting glucose molecule) are generated in the mitochondrial matrix.

Several recent in vitro and in vivo observations provide evidence for an increase in alanine and succinate in brain, muscle, and heart of hypoxic mammals (for references see Sanborn et al. 1979; Freminet 1981). In conclusion of their study on rats exposed for 28 days at the simulated altitude of 7 km in a hypobaric chamber, Cascarano et al. (1976) suggested that the succinate produced in peripheral hypoxic tissues is transported by the blood to the more oxygenated lungs; there it might be oxidized to fumarate and malate via the second span of the normally operating Krebs cycle.

For the effects of hypoxia on the brain concentration of amino acids, much contradictory data have been reported. Discarding the experiments made under acute conditions in anesthetized animals, the following remains. Unanesthetized rats were kept by Weyne et al. (1977) in normobaric hypoxia equivalent to 6.0–8.0 km altitude for 2 h to 7 days. Apart from a threefold increase in lactate concentration, glutamate and aspartate concentrations were found to be less in the hypoxic brains than in controls, whereas that of γ-aminobutyrate (GABA) was higher. When hypocapnia was prevented by adding CO_2 to the inhaled gas mixture, only the increase in brain GABA was observed, thus indicating that other changes were largely hypocapnia-dependent.

On a weight unit basis oxidation of fat yields 2.3 times more energy than that of carbohydrate oxidation. The process, however, requires 2.4 times more oxygen. Hence, in hypoxia, the utility of fatty acids as a fuel for the Krebs cycle can be questioned. Unfortunately, available information on the fatty acid metabolism at high altitude is scanty. In rats exposed for 7 days at a simulated altitude of 7 km in a hypobaric chamber, Kinnula and Hassinen (1978) found impaired capacity for oxidation of fatty acid by mitochondria of liver and heart, and accumulation of triglycerides in the liver. On the other hand, after returning from a 4 weeks' stay at altitudes ranging from 5.5 to 7.0 km in the Himalayas, the myocardial utilization of free fatty acids (FFA) was reported to decrease, and that of lactate to increase (Moret et al. 1981). Altogether, these observations suggest that FFA might account for a smaller fraction of oxidative metabolism at high altitude. Hypoxic hyperventilation is associated with increased plasma lactate concentration (Sect. 6.2.5), and the accumulation of lactate in blood appears to have inhibitory effects on FFA release (Issekutz et al. 1975). However, catecholamines may overcome these effects and, under hypoxia, could facilitate both the mobilization and utilization of FFA (Jones et al. 1972). Capacities for fat oxidation appear to be high in the high altitude camelids (Hochachka et al. 1983).

6.2.5 Glycolysis

Since the oxidative conversion from glucose to pyruvate and lactate does not require oxygen and produces some ATP (Figs. 6.1 and 6.2), glycolysis is accepted as the primeval mechanism for providing energy supply under anaerobic condi-

Table 6.2. Comparative value of the energy yielded anaerobically in mammals and reptiles

		Mouse	Lizard	Lizard/Mouse
1 Oxygen consumption	mmol/min	0.041	0.005	0.12
2 Body temperature	°C	36.8	37.0	1
3 Body mass	g	34	32.0	0.94
4 ATP production in aerobiosis	mmol/min	0.25	0.03	0.12
5 Muscle glycogen	g	0.04	0.04	1
6 ATP yield from anaerobic glycogen breakdown	mmol	0.72	0.72	1
7 Useful time of anaerobiosis	min	3	24	8

The data of lines 1 to 3 have been obtained on mice *Mus musculus* at 32°–33 °C ambient temperature, and on the lizard *Amphibolurus nuchalis* kept at 37 °C (Else and Hulbert 1981); note that body mass and body temperatures are comparable in the two vertebrates. The calculations in lines 4 to 7 are derived as follows. (See Coulson et al. 1977). Line 4 = from line 1, assuming 6 ATP molecules are produced for each O_2 molecule consumed; line 5 = from line 3, assuming that the skeletal muscle mass is 1/4 of the total body mass, and contains 5‰ of glycogen; line 6 = from line 5, assuming that each gram of glycogen corresponds roughly to 6 mmol, expressed as glucose, and that complete conversion of 1 glycogen molecule to lactate yields 3 ATP molecules; line 7 = line 6 divided by line 4.

Glycolysis by which all muscle glycogen would be converted to lactate would provide energy at the rate of the energy requirement (line 4) for a much longer time in the reptile than in the mammal. The difference would be reinforced by allowing the body temperature to decrease in the reptile, thus decreasing its energy requirement

tions. Yet the energy derived from glycolysis is only 1/18 of that derived from the full oxidation of glucose (Fig. 6.2). On the other hand, as shown in Table 6.2 (line 7), the energy yield anaerobically may be of different effectiveness among vertebrates; the lower the metabolic rate, the greater the fraction of total energy produced that may be derived from glycolysis (Coulson et al. 1977).

a) Increased lactate concentrations in blood and tissues is a common, nondisputed finding in man and mammals translocated from low to high altitudes. Thus, in the rat liver studied by Reed and Pace (1980) at the altitude of 3.8 km (Fig. 6.4 A), the lactate concentration roughly doubled within the first hour of altitude exposure, declined thereafter, but remained significantly different from controls after 20–60 days. Similar changes occurred in the lactate/pyruvate ratio (Fig. 6.4 C). Another example, concerned with heart, is provided in Fig. 6.7, which summarized some results obtained by Vergnes (1973) on rats translocated from near sea level to the altitude of 3.5 km.

In brain, the increase in lactate concentration is the earliest biochemical change observed during hypoxia (Fig. 6.8, bottom), while the energy status remains unaffected (Cohen 1973; Duffy et al. 1972; Siesjö 1977; Weyne et al. 1977). The hypocapnia provoked by hypoxic hyperventilation plays an important role in the increased lactate concentration (Cohen 1972); giving some CO_2 to breathe limits anaerobic glycolysis during hypoxia (Weyne et al. 1977).

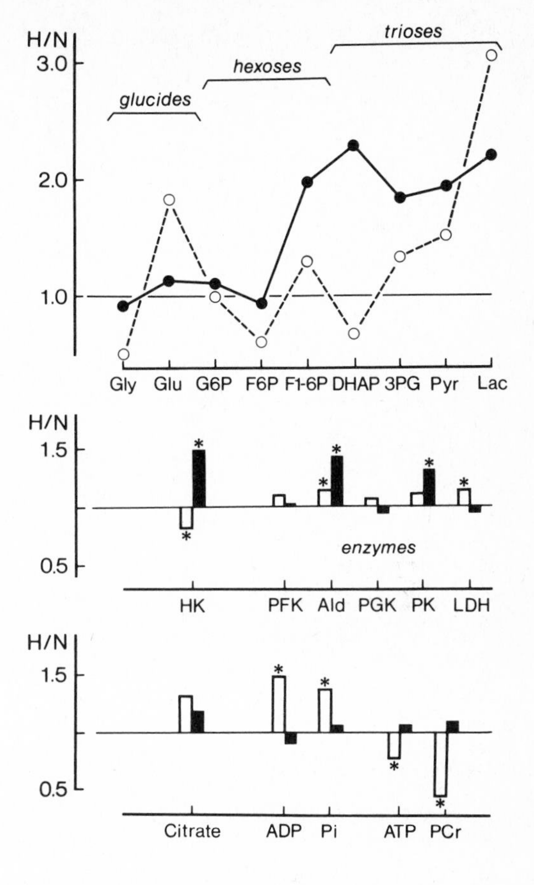

Fig. 6.7. Relative changes from controls of glycolytic metabolites (*top*) and some related enzymes (*middle*) in rat heart in the course of 85 days' exposure to 3.5 km in the Alps (data from Vergnes 1973). In the *lower part* the relative changes in the concentrations of citrate and adenosine phosphate system components are shown. *Open circles and columns:* data after 3 days at high altitude; *closed circles and columns:* data after 12 weeks at high altitude. *Stars* significant difference from control

As far as muscle is concerned, acute hypoxia does not reduce the formation of lactate during exercise, whereas chronic exposure to hypoxia causes a reduction in the amount of lactate formed. Thus, in native highlanders, the accumulation of blood lactate during exercise is lower than in sea-level subjects (Hurtado 1964). This is suggestive of an aerobic rather than anaerobic source of energy in high-altitude species.

Among lower vertebrates, little is known about the anaerobic metabolism under altitude hypoxia. Blood lactate concentration has been reported to increase in fish made hypoxic, and to return to normal after about 1 month of acclimation (Greaney et al. 1980). No indications of altitude adaptation were found in sceloporine lizards, when the duration of maximal activity and the amount of lactate formed were compared in lowland and in highland animals (Bennett and Ruben 1975).

b) Lactate dehydrogenase (LDH) is the enzyme that catalyses the reduction of pyruvate to lactate, thereby allowing the cytosolic NAD^+ to be regenerated (Fig. 6.2). Due to its dual function in the glycolytic sequence and redox balance of the cytosol, LDH has received much attention; but the reported data are conflicting.

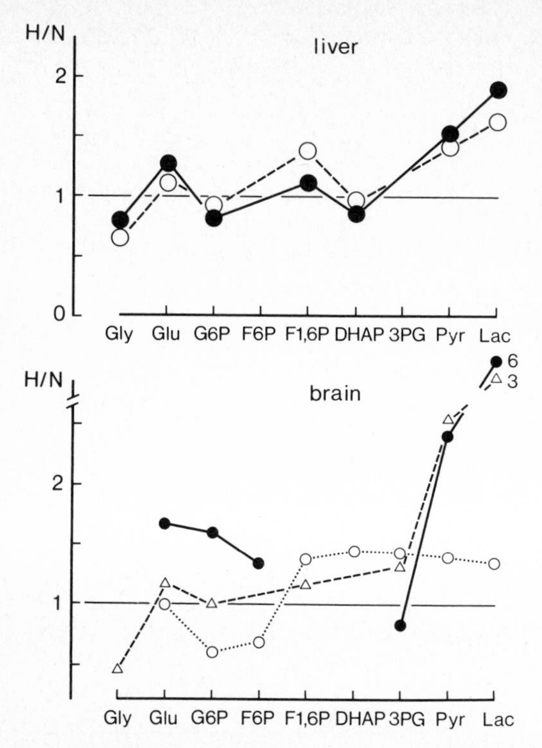

Fig. 6.8. Relative changes from controls of glycolytic metabolites in liver (*above*) and brain (*below*) in the course of hypoxia. *Top: open* and *closed circles* refer to data obtained in the rat after 3 and 63 days of exposure to 3.8 km altitude on White Mountain. (After Cipriano and Pace 1973). *Bottom: triangles* refer to data obtained in conscious mice exposed for 15 min to normobaric hypocapnic hypoxia equivalent to 10.5 km altitude. (After Duffy et al. 1972). *Open* and *closed circles* refer to data obtained in anesthetized and artificially ventilated rats exposed for 1 and 30 min, respectively, to normobaric hypoxia such as arterial P_{O_2} decreased to about 25 Torr. (After Norberg and Siesjö 1975)

In the brain, Miller and Hale (1968) found no difference between Sprague-Dawley rats exposed for 2 months at 5.5 km in an altitude chamber and control animals, either in LDH activity or in the relative proportions of the various isozymes. Other tissues (heart, liver, diaphragm, and gastrocnemius), similarly, showed no change. By contrast, in the brain of Wistar rats (hence a different strain) exposed for only 7 days (a shorter time) to normobaric hypoxia (a different type of hypoxic exposure) equivalent to 5.7 km, Berlet et al. (1981) found an increased proportion of H-type isozymes, thought to be related with aerobic metabolism, and an increase in LDH activity. Conversely, a higher concentration of M-subunits, thought to be associated with anaerobic metabolism, has been reported in the brain of humans who were exposed to 3.8 km for their entire life, in comparison with humans living at sea level (Hellung-Larsen et al. 1973).

In cardiac muscle, one is faced with similarly opposing views. In short, Harris et al. (1970) found a higher LDH activity only in guinea pigs, but not in rabbits and dogs native to 4.3 km, than in sea-level animals. Penney (1974) detected isozyme variations in LDH in rat heart, whereas Vergnes et al. (1976) did not. The altitude, however, was lower in the second study (3.5 km instead of 6.2 km). At an intermediate altitude (4.4 km), Mager et al. (1968) found no change in the LDH activity, but an increased proportion of M-subunits. Dietary restriction and cardiac hypertrophy appear to be important in the modulation of LDH activity (Barrie and Harris 1976, Table 6.3). In a recent study, Hochachka et al. (1983) found that the LDH activity of the left ventricle was lower in high-altitude llama

Table 6.3. Summary of the significant changes in some myocardial enzyme activities in guinea pigs exposed to the altitude of 5 km in a hypobaric chamber for 28 days (derived from Barrie and Harris 1976)

Metabolic pathway	Enzymes	Diet restriction	Myocardial hypertrophy	Chronic hypoxia
Glycogen mobilization	Glycogen phosphorylase	↓	↑	
	Phosphoglucomutase		↑	
Hexoses	Hexokinase	↓	↑	↓
	Glucose phosphate isomerase		↑	
	Phosphofructokinase			
	Aldolase			↑
Pentoses (shunt)	G6P dehydrogenase			
Trioses	G3P dehydrogenase		↑	↑
	Phosphoglycerate kinase		↑	↑
	Pyruvate kinase			↑
	Lactate dehydrogenase	↑	↑	
Fatty acids	α-hydroxybutyrate dehydrogenase			↑
Krebs cycle	Succinate dehydrogenase			Transient ↑

Arrows visualize the changes attributable to the effects of diet restriction, myocardial hypertrophy or/and chronic hypoxia

and alpaca than in the sea-level ox, an observation suggesting that high-altitude species could display a reduced dependence upon anaerobic glycolysis.

c) Glycogen and glucose concentrations in liver, heart, and muscle have been shown to increase, decrease, or remain unchanged during altitude hypoxia, depending on the experimental conditions (Timiras et al. 1958; Ou 1974). Concerning the heart, a most common finding in moderate hypoxia is illustrated in Fig. 6.7 (left upper part). There is evidence that glucose utilization was reduced in the early phase of hypoxia. Similar observations have been made for liver and brain (Cipriano and Pace 1973; Norberg and Siesjö 1975).

d) Other glycolytic intermediates and enzymes have been little studied under altitude hypoxia. Their relative changes from control in various tissues of rats exposed for 3–85 days at 3.5–3.8 km altitude are shown in Figs. 6.7 and 6.8.

These figures suggest the existence of a first regulatory step from G6P to F1,6P, through the action of phosphofructokinase (PFK). Current views regard PFK as the major pacemaker for glycolysis; also, there is evidence that glycolytic flux is stimulated when the intracellular pH increases, as is the case at least transiently during hypocapnic hypoxia (Fidelman et al. 1982).

Another regulatory step can be identified further along the glycolytic sequence, i.e., that leading to pyruvate through the action of pyruvate kinase (PK). In the course of prolonged stay at high altitude, the activity of PK was found to increase in the heart of rat (Fig. 6.7) and guinea pigs. In the latter, such a change appears to be specifically associated with hypoxia (Barrie and Harris 1976, Table 6.3).

e) Activity of the pentophosphate shunt (PP at the G6P branching point in Fig. 6.2, top) has been reported to be reduced in the liver of rats exposed for 63 days to 3.8 km at the Barcroft Laboratory (Cipriano and Pace 1973). This finding is in agreement with the conclusion of Blume and Pace (1957) based on the kinetics of ^{14}C-labeled glucose oxidation in the whole mouse exposed at the same altitude. By contrast, the pentose pathway has been found to be activated, in the heart of rats exposed for 85 days at a similar altitude in the Alps (Vergnes 1973) and in that of guinea pigs exposed to 5 km for 28 days in a hypobaric chamber (Barrie and Harris 1976).

6.3 Nonbioenergetic Adaptations

Not only is the synthesis of important organic compounds tightly linked to the aerobic degradation of substrates, but molecular oxygen is also used in reactions that are largely extramitochondrial, and do not yield ATP (Bloch 1962). In recent years, several reports have shown that hypoxia induces alteration in these nonenergonic reactions. Some of these reports are considered under the following subheadings (1) neurotransmitters, (2) vasoactive metabolites, (3) nucleic acids.

6.3.1 Neurotransmitters

Hypoxia is commonly invoked to explain alterations in mental function. These alterations are thought to be caused by "transmission failure" rather than "energy failure" (Siesjö 1978). For instance, when human beings are acutely exposed to hypobaric hypoxia, delay in dark adaptation occurs at a simulated altitude close to 1.2 km (McFarland and Evans 1939). At slightly higher altitude, there is impaired ability to learn a complex task. Short memory is affected at about 3.7 km, and critical judgement may be lost above 5 km. Such impairments of brain function may relate to alteration in neurotransmitter metabolism.

6.3.1.1 Monoamines

Figure 6.9 summarizes the main pathways whereby dopamine, norepinephrine, and serotonin are formed and degraded. Note that the enzymes tyrosine hydroxylase and tryptophan hydroxylase, presumed to constitute the rate-limiting steps in the synthesis of catecholamines and serotonin, respectively, require molecular oxygen for the synthesis of DOPA (dihydroxyphenylalanine) and 5-HTP (5-hydroxytryptophan), respectively. Original references and comprehensive discussion of this topic are available in Siesjö (1978).

The hydroxylation of tyrosine and tryptophan can be estimated using NSD-1015 (3-hydroxybenzyl hydrazine), a potent inhibitor of the amino acid decarboxylase. This leads to the accumulation of any DOPA formed from the hydroxylation of tyrosine, or of any 5-HTP formed from the hydroxylation of tryptophan.

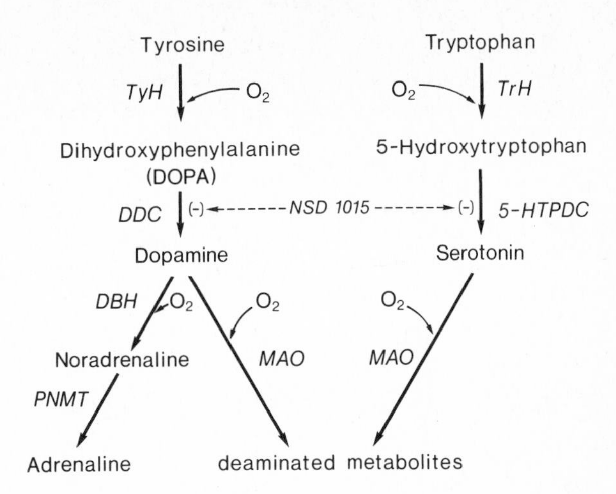

Fig. 6.9. Biosyntheses of catecholamines from tyrosine, and of serotonin from tryptophan. Schematic pathways showing the utilization of oxygen and the enzymes involved: tyrosine hydroxylase (TyH), DOPA decarboxylase (DDC), dopamine-β-hydroxylase (DBH), phenylethanolamine N-methyltransferase (PNMT), monoamine oxidase (MAO), tryptophan hydroxylase (TrH), 5-hydroxytryptophan hydroxylase (5-HTPDC). Also indicated is NSD-1015 (3-hydroxybenzyl hydrazine), which blocks the enzymes DDC and 5-HTPDC

a) In vivo experiments suggest that the hydroxylation of both tyrosine and tryptophan is sensitive to the availability of O_2 within the physiological range of P_{O_2}. Indeed, after inhibition of decarboxylase with NSD-1015, the accumulation of DOPA and 5-HTP in brain was found to decrease linearly with inspired P_{O_2} in conscious, adult rats exposed for 30 min to various levels of normobaric hypoxia (Davis and Carlsson 1973). When the hypoxia was prolonged, however, synthesis of monoamine in the brain returned to control levels within about 1 day (Davis 1975), indicating that some adaptive process in the O_2 delivery system occurred. Hedner (1978), Hedner et al. (1977), and Hedner and Lundborg (1979) made similar observations in newborn and adolescent rats. Hughes et al. (1983) found the norepinephrine levels of the midbrain to be lower in rabbits acclimatized to hypobaric hypoxia (simulated altitude of 6 km on alternate days for 70 days) than in controls; but rats acclimatized in the same manner did not differ from controls.

There are observations indicating that alterations in behavioral and locomotor activity under hypoxia (Brown and Engel 1973) are due to the acute decrease in brain monoamine synthesis. Even the suprapontine respiratory control system may be affected (Gautier and Bonora 1980). Preventive administration of DOPA greatly reduces these deleterious effects of hypoxia on behavior and motor function in rats (Brown and Engel 1973; Boismare et al. 1977; Boismare 1979), as well as the tachypnea induced by hypoxia in the carotid body-denervated cats (Gautier and Bonora 1980). Dopaminergic agonists (apomorphine, bromocriptine, amantidine, piribedil) similarly oppose the effects of hypobaric hypoxia on behavior and learning in rats (Saligaut et al. 1981). In these studies, however, improvement of O_2 delivery via a side effect on cerebral circulation could not be excluded (Harper and McCulloch 1977).

For adrenal catecholamines, acclimatization of rats for 1 year in natural conditions at the altitude of 1.35 km has been reported to result in an activity of plasma dopamine-β-hydroxylase (Fig. 6.9) twice as high as in control animals maintained at 0.15 km (Balaz et al. 1980). Furthermore, when the rats were submitted to an "immobilization stress," the observed changes in the norepinephrine concentration, and in the activities of dopamine-β-hydroxylase and tyrosine hydroxylase in the adrenal gland and plasma indicated a more intensive activation of the sympathetic adrenomedullary system in the acclimatized animals, compared with control ones. According to Davis (1976), during neuronal stimulation, an allosteric change in tyrosine hydroxylase may increase the affinity of the enzyme for oxygen, thereby allowing greater catecholamine synthesis despite limiting O_2 availability.

b) In vitro experiments, although not safeguarded from potential errors due to the existence of diffusional gradients of O_2 tension, provide arguments in favor of the dependence of monoamine metabolism on O_2. Tyrosine hydroxylase has a relatively high affinity for oxygen in isolated synaptosomes (P_{50}, the O_2 partial pressure required for the reaction rate to be half maximal, is about 12 Torr; Davis 1977), suggesting that the reaction might be retarded, even with moderate O_2 depletion, since O_2 tensions in brain tissue range between 10 and 20 Torr in normoxia (Smith et al. 1977). The same may be true for tryptophan hydroxylase reaction ($P_{50}=3$–10 Torr, depending on the tryptophan concentration; Katz 1980) and, a fortiori, for the monoamine oxidase reaction (MAO in Fig. 6.9; $P_{50}=100$ Torr; Roth 1979). Furthermore, in hypothalamic synaptosomes, the dopamine-β-hydroxylase reaction appears not to be saturated with respect to oxygen, even under normoxic values of P_{O_2}; an increment of 30–40% is observed when assays are performed under O_2 at one atmosphere (Katz 1982).

These studies suggest that monoamine metabolism can be modulated by changes in P_{O_2} within the physiological range. However, since both synthesis and degradation of norepinephrine, dopamine, and serotonin are O_2-dependent, it is not possible to predict what exactly is the effect of hypoxia on the turnover of these compounds in vivo. Also, aging processes may significantly affect monoamine mechanisms (Robinson 1980). On the other hand, the oxygen affinities of the oxygenases vary with the conditions of assay, and with the concentrations of substrates or activators (Kaufman and Fisher 1974; Tipton 1972). In particular, the affinity of purified dopamine-β-hydroxylase for oxygen depends upon the concentrations of dopamine, ascorbate, and fumarate (Goldstein et al. 1968). The absence of information on local concentrations of these metabolites impairs conclusions from kinetic studies on isolated enzymes with respect to the oxygen dependence of amine synthesis and catabolism in tissues. At any rate "... some pathways which regulate transmitter metabolism are affected at reductions in Pa_{O_2} values that are associated with changes in functional behavior" (Siesjö 1978, p 434).

6.3.1.2 Acetylcholine

Acetylcholine is an important neurotransmitter, the synthesis of which does not require molecular oxygen directly, but is nevertheless tightly linked to oxidative metabolism. As the well-known reaction

$$\text{choline} + \text{acetyl-CoA} \rightarrow \text{acetylcholine} + \text{CoA}$$

shows, energy is needed in the form of acetyl-CoA, which results from the oxidation of pyruvate (Fig. 6.2). Not only must acetyl-CoA be formed, but it must also be transported out of the mitochondrion (by an unknown acetyl carrier) into the cytosol, where it combines with choline through the action of the enzyme choline-acetyl transferase. Normally, synthesis of ACh in brain occurs at a rate corresponding to about 10% of the maximal enzyme activity, with less than 1% of total pyruvate oxidation being diverted to ACh synthesis (see Siesjö 1978). Several studies, however, indicate that mild depression of O_2 supply reduces synthesis of ACh, with functional consequences.

Gibson et al. (1975) reported first that inhibition of the oxidation of pyruvate proportionally reduced ACh synthesis in brain slices. In vivo, they demonstrated that histotoxic hypoxia (KCN), as well as anemic hypoxia ($NaNO_2$), impair synthesis of ACh from labeled precursors in rat brain (Gibson and Blass 1976). In mice made similarly hypoxic with cyanide or nitrite, a double-label technique using [U-^{14}C]-glucose and [^{3}H]-choline permitted Gibson et al. (1978) to demonstrate a 43% decrease in the synthesis of ACh in the forebrain without significant changes in cerebral lactate or cyclic AMP. More recently, normobaric hypoxia was acutely tested in unanesthetized rats, using the same double-label technique (Gibson and Duffy 1981; Gibson et al. 1981 a, b). A strong decrease in ACh synthesis was observed in the hypoxic animals (in which the arterial P_{O_2} was 57 or 42 Torr), compared with those made slightly hyperoxic ($Pa_{O_2}=120$ Torr). Pretreatment with physostigmine reduced the rate of ACh synthesis necessary to maintain ACh levels, and protected animals against the effects of hypoxia (Gibson and Blass 1976). Whether chronic hypoxia yields adaptive response subserving more effective ACh synthesis is not known.

Since the effects of a variety of neurotransmitters and hormones are mediated through cyclic AMP or cyclic GMP, one may wonder whether changes in these cyclic nucleotides occur in the course of hypoxia. The data of Gibson et al. (1978) raise the possibility that the cyclic GMP increases in the brain of mice treated either with injections of $NaNO_2$ (anemic hypoxia) or KCN injections (histotoxic hypoxia). On the other hand, under severe, acute normobaric hypoxia ($Pa_{O_2}=18$–22 Torr for 15–30 min), Siesjö and his associates found no change in the concentration of cyclic nucleotides (Folbergrova et al. 1981).

6.3.2 Vasoactive Metabolites

It has long been recognized that an increase in metabolism or a decrease in O_2 availability in certain systemic tissues causes local vasodilation. Also, there is evidence indicating a role of metabolically linked compounds in local regulation of blood flow (Haddy and Scott 1968). Among these are adenosine and prostaglandins.

6.3.2.1 Adenosine

Interest in the regulatory functions of adenosine arose primarily from studies by Berne and associates, which suggested that this nucleotide might function as physiological regulator of coronary blood flow (Berne 1963). This concept was extended later to regulation of the vascular beds of kidney, skeletal muscle, and brain (Berne et al. 1974). It has been confirmed that adenosine is a potent vasodilator, which is supplied readily from intracellular nucleotide pool (mainly via dephosphorylation of AMP), is transported across the cell membrane, appears in the extracellular fluid, and is subsequently metabolized in inactive degradation products (inosine and hypoxanthine). Multiple pathways and mechanisms appear to be involved in adenosine metabolism (Siesjö 1978; Winn et al. 1981).

Several procedures known to influence cerebral blood flow were undertaken in the rat to determine if they altered the adenosine concentration in brain. For hypoxia, only transient exposures (<5 min) in normobaric conditions have been tested in anesthetized, artificially ventilated animals. Results are disparate. An increase in the brain concentration of adenosine has been reported by Rubio et al. (1975) and Winn et al. (1981), when decreasing the arterial P_{O_2} to about 30 Torr. Rehncrona et al. (1978), on the other hand, failed to confirm such an observation.

6.3.2.2 Prostaglandins

Prostaglandins are released by most animal tissues from membrane phospholipids. They require oxygen for their synthesis, when the precursor fatty acid (principally arachidonic acid) is oxygenated through the action of the enzyme cyclo-oxygenase. Endoperoxides result, which are converted enzymatically to a variety of prostaglandins (PG's; the exact products are dependent upon the specific enzymes at work in the tissue under consideration). The question of the O_2 requirements for synthesis of PG's is presently unresolved.

While the mechanisms by which PG's may stimulate or inhibit smooth muscle activity is yet not clear (Bolton 1979), there is evidence indicating that PG's may contribute to functional vasodilatation in a variety of tissues (Kaley 1978). Increased synthesis of PG's, vasodilatation and increased coronary blood flow have been reported to occur when the O_2 tension was caused to decrease from 100 to 38 Torr (Kalsner 1977). The following hypothesis has been proposed by Markelonis and Garbus (1975): when the capacity to phosphorylate ADP to ATP decreases in hypoxic cells, the sequestered calcium is released and activates the enzyme phospholipase A_2, which releases the precursor fatty acid from phospholipids, thereby yielding substrate induction for synthesis of PG's. On the other hand, calcium has also been shown to be an important regulator of synthesis of PG's (Rubin and Laychock 1978).

Since prostaglandins, most probably present in endothelia of all blood vessels, not only influence local blood flow but also have a profound effect on vascular permeability (Kaley 1978), it might be relevant to study their contribution in the process of adaptation to altitude hypoxia. Under severe hypobaric hypoxia (equivalent altitude = 9.7 km), pretreatment with prostacyclin was found to increase considerably the survival time of mice (Nikolov et al. 1982).

6.3.3 Nucleic Acids

The syntheses of proteins are achieved through the action of DNA (desoxyribonucleic acid) and RNA (ribonucleic acid) molecules. Whether or not these molecules are affected under altitude hypoxia has been little studied.

One current view, advocated by Meerson (1975), holds that a marked activation of the synthesis of nucleic acids and proteins occurs in the process of adaptation to intermittent hypobaric hypoxia (simulated altitude = 6–7 km, 6 h daily for 40 days). Found not only in the hypertrophied heart, but also in the cerebral cortex of rats translocated from sea level, this activation was thought to yield an increase in the capacity of the respiratory chain (Sect. 6.2.3.2 d). Others, however, under roughly similar conditions, found either no change (Stere and Anthony 1977) or decreased synthesis and concentration of DNA, RNA, and proteins (Serra et al. 1981; Trojan et al. 1978) in the rat brain and heart. In lungs, chronic hypoxia has been reported to stimulate DNA synthesis and proliferation of various cell types. Thus, autoradiographic investigations showed an increase in the [^{3}H]-thymidine labeling index of cells in the bronchial epithelial lining, of parenchyma cells, and of smooth muscle cells in the media of pulmonary arteries after 9 days of exposure to the simulated altitude of 4.25 km in a hypobaric chamber (Niedenzu et al. 1981). Apparently there are no studies on native high-altitude animals.

6.4 Concluding Remark

The data here presented make it evident that the biochemical adjustments to altitude hypoxia are poorly understood to date. No definite conclusion can be drawn from the contradictory observations accumulated in the past decades. Many differences in experimental situations and protocols (some of which have been detailed in the introduction of this chapter) can be invoked to explain the differences in results. If, 20 years after Barbashova, we can still say that "it is hoped that through the combined efforts of specialists in different fields, new data will be obtained which will make it possible to gain a deeper understanding of the nature of adaptation to hypoxia at the cellular level" (Barbashova 1964, p 50), it is highly desirable that well-controlled studies account coherently for all known factors involved.

Accepting then, for the present, only such adaptations as have been clearly demonstrated to serve in the compensation of altitude hypoxia, one may rank them as follows: (1) increased ventilation, which raises the O_2 partial pressure in the external gas exchange organ above what it would normally be, and (2) increased blood O_2 capacitance, which limits the decrease of P_{O_2} in the systemic capillaries. Thus, the conclusion of an early review on the physiological effects of altitude hypoxia (Schneider 1921) remains valid.

Appendix

Table A.1. Barometric pressure (P_B) and partial pressure of oxygen in dry air (P_{O_2}) vs. altitude

Altitude		P_B				P_{O_2}	Equivalent F_{O_2} at
Meter	Foot	kPa	Torr	atm	mb	Torr	P_B = 760 Torr
0	0	101.3	760	1.000	1013	159	0.2095
500	1640	95.5	716	0.942	955	150	0.197
1000	3281	89.9	674	0.887	899	141	0.186
1500	4922	84.6	634	0.835	846	133	0.176
2000	6562	79.5	596	0.785	795	125	0.164
2500	8202	74.7	560	0.737	747	117	0.154
3000	9842	70.1	526	0.692	701	110	0.145
3500	11,483	65.7	493	0.649	658	103	0.136
4000	13,123	61.6	462	0.608	616	97	0.128
4500	14,764	57.7	433	0.570	577	91	0.120
5000	16,404	54.0	405	0.533	540	85	0.112
5500	18,045	50.5	379	0.498	505	79	0.104
6000	19,685	47.2	354	0.466	472	74	0.097
6500	21,325	44.0	330	0.435	440	69	0.090
7000	22,966	41.1	308	0.405	411	65	0.085
7500	24,606	38.3	287	0.378	382	60	0.079
8000	26,247	35.6	267	0.351	356	56	0.074
8500	27,887	33.1	248	0.327	331	52	0.068
9000	29,527	30.8	231	0.303	307	48	0.064

P_B values are from the Manual of the ICAO Standard Atmosphere (International Civil Aviation Organization 1964). P_{O_2} values are equal to the product $0.2095 \cdot P_B$, where 0.2095 is the fractional concentration of oxygen (F_{O_2}) in the normal ambient air. The last column shows the F_{O_2} values (calculated as $P_{O_2}/760$) which, at sea level, would yield the value of P_{O_2} prevailing in dry air at the various altitudes

Table A.2. Saturation pressure of water vapor (P_{H_2O}) at various temperatures (T)

T °C	P_{H_2O} Torr	T °C	P_{H_2O} Torr	T °C	P_{H_2O} Torr
−30	0.3	10	9.2	30	31.8
−25	0.5	11	9.8	31	33.7
−20	0.8	12	10.5	32	35.7
−15	1.2	13	11.2	33	37.7
−10	2.0	14	12.0	34	39.9
− 5	3.0	15	12.8	35	42.2
− 4	3.3	16	13.6	36	44.6
− 3	3.6	17	14.5	37	47.1
− 2	3.9	18	15.5	38	49.7
− 1	4.2	19	16.5	39	52.4
0	4.6	20	17.5	40	55.3
1	4.9	21	18.7	41	58.3
2	5.3	22	19.8	42	61.5
3	5.7	23	21.1	43	64.8
4	6.1	24	22.4	44	68.2
5	6.5	25	23.8	45	71.8
6	7.0	26	25.2	46	75.6
7	7.5	27	26.7	47	79.6
8	8.0	28	28.3	48	83.7
9	8.6	29	30.0	49	88.0

Table A.3. O_2 and CO_2 capacitances ($\mu mol \cdot L^{-1} \cdot Torr^{-1}$) in distilled water ($\beta w_{O_2}$ and βw_{CO_2}), and in air ($\beta g_{O_2} = \beta g_{CO_2}$) at various temperatures (T, °C). (Dejours 1981)

T	βw_{O_2}	βw_{CO_2}	$\beta g_{O_2} = \beta g_{CO_2}$
0	2.88	102.0	58.7
10	2.24	70.6	56.6
20	1.82	51.4	54.7
30	1.54	39.2	52.9
40	1.36	31.2	51.2

References

Acker H, Fidone S, Pallot D, Eyzaguirre C, Lübbers DW, Torrance RW (eds) (1977) Chemoreception in the carotid body. Springer, Berlin Heidelberg New York, p 296

Adachi H, Strauss HW, Ochi H, Wagner HJ (1976) The effect of hypoxia on the regional distribution of cardiac output in the dog. Circ Res 39:314–319

Adams WE (1958) The comparative morphology of the carotid body and carotid sinus. Thomas, Springfield, Ill, p 272

Adams WH, Strang LJ (1975) Hemoglobin levels in persons of Tibetan ancestry living at high altitude. Proc Soc Exp Biol Med 149:1036–1039

Adamson JW, Finch CA (1975) Hemoglobin function, oxygen affinity, and erythropoietin. Annu Rev Physiol 37:351–369

Adolph EF (1956) General and specific characteristics of physiological adaptations. Am J Physiol 184:18–28

Adolph EF (1964) Perspectives of adaptation: some general properties. In: Dill DB, Adolph EF, Wilber CG (eds) Handbook of adaptation to the environment. Am Physiol Soc, Washington, DC, pp 27–35

Albaum HG, Chinn HI (1953) Brain metabolism during acclimatization to high altitude. Am J Physiol 174:141–145

Altland PD, Rattner BA (1981) Age and altitude tolerance in rats: temperature, plasma enzymes, and corticosterone. J Appl Physiol: Respir Environ Ex Physiol 50:367–373

Altman PL, Dittmer DS (eds) (1974) Biology data book, 2nd edn. Fed Am Soc Exp Biol, Bethesda, MD, pp 1876–1897

Anthony A, Ackerman E, Strother GK (1959) Effects of altitude acclimatization on rat myoglobin. Changes in myoglobin concentration of skeletal and cardiac muscle. Am J Physiol 196:512–516

Arieli R, Ar A (1979) Ventilation of a fossorial mammal (*Spalax ehrenbergi*) in hypoxic and hypercapnic conditions. J Appl Physiol: Respir Environ Ex Physiol 47:1011–10

Arieli R, Ar A, Shkolnik A (1977) Metabolic responses of a fossorial rodent (*Spalax ehrenbergi*) to simulated burrow conditions. Physiol Zool 50:61–75

Asmussen E, Chiodi H (1941) The effect of hypoxemia on ventilation and circulation in man. Am J Physiol 132:426–436

Asmussen E, Nielsen M (1955) The cardiac output in rest and work at low and high oxygen pressures. Acta Physiol Scand 35:73–83

Asmussen E, Nielsen M (1960) Alveolar-arterial gas exchange at rest and during work at different O_2 tensions. Acta Physiol Scand 50:153–166

Aste-Salazar J, Hurtado A (1944) The affinity of hemoglobin for oxygen at sea level and at high altitude. Am J Physiol 142:733–743

Åstrand PO, Rodahl K (1970) Textbook of work physiology. McGraw-Hill, New York, p 669

Baker PT (ed) (1978) The biology of high-altitude peoples. Cambridge Univ Press, Cambridge, p 357

Baker PT, Little MA (1976) Man in the Andes. A multidisciplinary study of high-altitude Quechua. Dowden, Hutchinson & Ross, Strondsburg, Pa, p 482

Balaz J, Balazova E, Blazicek P, Kvetnansky R (1980) The effect of one-year acclimatization of rats to mountain conditions on plasma catecholamines and dopamine-β-hydroxylase activity. In: Usdin E, Kvetnansky R, Kopin IJ (eds) Catecholamines and stress: recent advances. Elsevier/North-Holland, Amsterdam, pp 259–264

Balke B (1964) Cardiac performance in relation to altitude. Am J Cardiol 14:796–810

Balke B (1972) Physiology of respiration at altitude. In: Yousef MK, Horwath SM, Bullard RW (eds) Physiological adaptations, desert and mountain. Academic Press, London New York, pp 195–208

Ballintijn CM (1972) Efficiency, mechanics and motor control of fish respiration. Respir Physiol 14:125–141

Bamford OS (1974) Oxygen reception in the rainbow trout (*Salmo gairdneri*). Comp Biochem Physiol 48A:69–76

Banchero N (1975) Capillary density of skeletal muscle in dogs exposed to simulated altitude. Proc Soc Exp Biol Med 148:435–439

Banchero N (1982) Long-term adaptation of skeletal muscle capillarity. Physiologist 25:385–389

Banchero N, Grover RF (1972) Effect of different levels of simulated altitude on O_2 transport in llama and sheep. Am J Physiol 222:1239–1245

Banchero N, Grover RF, Will JA (1971) Oxygen transport in the llama (*Lama glama*). Respir Physiol 13:102–115

Banchero N, Gimenez M, Rostami A, Eby SH (1976) Effects of simulated altitude on O_2 transport in dogs. Respir Physiol 27:305–321

Bär T (1980) The vascular system of the cerebral cortex. Adv Anat Embryol Cell Biol 59:1–62

Barbashova ZI (1964) Cellular level of adaptation. In: Dill DB, Adolph EF, Wilber CG (eds) Handbook of physiology, section 4. Adaptation to the environment. Am Physiol Soc, Washington, DC, pp 37–54

Barcroft J (1925) The respiratory function of the blood. Part I: Lessons from high altitude. Cambridge Univ Press, London, p 207

Barcroft J, Binger CA, Bock AV, Doggart JH, Forbes HS, Harrop GA, Meakins JC, Redfield AC (1923) Observations upon the effect of high altitude on the physiological processes of the human body carried out in the Peruvian Andes chiefly at Cerro de Pasco. Philos Trans R Soc London Ser B211:351–480

Barer GR, Herget J, Sloan PJM, Suggett AJ (1978) The effect of acute and chronic hypoxia on thoracic gas volume in anaesthetized rats. J Physiol (London) 277:177–192

Barer GR, Bee D, Wach RA (1983) Contribution of polycythaemia to pulmonary hypertension in simulated high altitude in rats. J Physiol (London) 336:27–38

Barrie SE, Harris P (1976) Effects of chronic hypoxia and dietary restriction on myocardial enzyme activities. Am J Physiol 231:1308–1313

Barron DH, Metcalfe J, Meschia G, Huckabee W, Hellegers A, Prystowsky A (1964) Adaptations of pregnant ewes and their fetuses to high altitude. In: Weihe WH (ed) The physiological effects of high altitude. Pergamon Press, Oxford, pp 115–125

Bartels H (1970) Prenatal respiration. North-Holland Publ Comp, Amsterdam, p 187

Bartels H, Baumann R (1977) Respiratory function of hemoglobin. Int Rev Physiol 14:107–134

Bartels H, Hilpert P, Barbey K, Betke K, Riegel K, Lang EM, Metcalfe J (1963) Respiratory functions of blood of the yak, llama, camel, Dybowski deer, and African elephant. Am J Physiol 205:331–336

Bartels H, Bartels R, Rathschlag-Schaefer AM, Robbel H, Ludders S (1979) Acclimatization of newborn rats and guinea pigs to 3000–5000 m simulated altitudes. Respir Physiol 36:375–389

Bartlett D (1970) Postnatal growth of the mammalian lung: influence of low and high oxygen tensions. Respir Physiol 9:58–64

Bartlett D (1979) Effects of hypercapnia and hypoxia on laryngeal resistance to airflow. Respir Physiol 37:293–302

Bartlett D, Remmers JE (1971) Effects of high-altitude exposure on the lungs of young rats. Respir Physiol 13:116–125

Bashford CL, Barlow CH, Chance B, Haselgrove J, Sorge J (1982) Optical measurements of oxygen delivery and consumption in gerbil cerebral cortex. Am J Physiol 242 (Cell Physiol 11):C265–C271

Bauer C (1974) On the respiratory function of haemoglobin. Rev Physiol Biochem Pharmacol 70:1–31

Beattie J, Smith AH (1975) Metabolic adaptations of the chick embryo to chronic hypoxia. Am J Physiol 228:1346–1350

Becker EL, Schilling JA, Harvey RB (1957) Renal function in man acclimatized to high altitude. J Appl Physiol 10:79–80

Bellingham AJ (1972) The physiological significance of the Hill parameter "n". Scand J Haematol 9:552–556

Belmonte C, Pallot DJ, Acker H, Fidone S (eds) (1981) Arterial chemoreceptors. Leicester Univ Press, Leicester, p 532

Bencowitz HZ, Wagner PD, West JB (1982) Effect of change in P_{50} on exercise tolerance at high altitude: a theoretical study. J Appl Physiol: Respir Environ Ex Physiol 53:1487–1495

Bennett AF (1978) Activity metabolism of the lower vertebrates. Annu Rev Physiol 400:447–469

Bennett AF, Dawson WR (1976) Metabolism. In: Gans C, Dawson WR (eds) Biology of the reptilia, vol V. Academic Press, London, New York, pp 127–223

Bennett AF, Licht P (1972) Anaerobic metabolism during activity in lizards. J Comp Physiol 81:277–288

Bennett AF, Ruben J (1975) High-altitude adaptation and anaerobiosis in sceloporine lizards. Comp Biochem Physiol 50A:105–108

Berlet HH, Stefanovitch V, Volk B, Lehnert T, Franz M (1981) Chronic normobaric hypoxia and subcellular isoenzyme patterns of rat brain lactate dehydrogenase. In: Stefanovitch V (ed) Animal models and hypoxia. Pergamon Press, Oxford, pp 75–84

Berne RM (1963) Cardiac nucleotides in hypoxia: possible role in regulation of coronary blood flow. Am J Physiol 204:317–323

Berne RM, Rubio R, Curnish RR (1974) Release of adenosine from ischemic brain: effect on cerebral vascular resistance and incorporation into cerebral adenine nucleotides. Cir Res 35:262–271

Berry L, Beuzeville C, Krumdieck C (1957) Metabolic studies of guinea-pigs native to the high Andes and to the Peruvian coastal plain. (See Mensen de Silva E, Cazorla A 1973)

Bert P (1978) Barometric pressure: researches in experimental physiology. In: Hitchcock MA, Hitchcock FA (eds and transl). College Book Comp, Columbus, OH, repr Undersea Med Soc, Bethesda, MD, p 1055

Besch EL, Kadono H (1978) Cardiopulmonary responses to acute hypoxia in domestic fowl. In: Piiper J (ed) Respiratory function in birds: adult and embryonic. Springer, Berlin Heidelberg New York, pp 71–78

Besch EL, Burton RR, Smith AH (1971) Influence of chronic hypoxia on blood gas tensions and pH in domestic fowl. Am J Physiol 220:1379–1382

Betz E (1972) Cerebral blood flow: its measurement and regulation. Physiol Rev 52:595–630

Bidart Y, Drouet L, Durand J (1975) Débit sanguin dans le muscle squelettique chez les sujets résidant et transplantés en altitude (3800 m). J Physiol (Paris) 70:333–337

Biscoe TJ (1971) Carotid body: structure and function. Physiol Rev 51:437–495

Biscoe TJ, Willshaw P (1981) Stimulus-response relationships of the peripheral arterial chemoreceptors. In: Hornbein TF (ed) Regulation of breathing. Dekker, New York, pp 321–345

Bisgard GE, Forster HV, Orr JA, Buss DD, Rawlings CA, Rasmussen B (1976) Hypoventilation in ponies after carotid body denervation. J Appl Physiol 40:184–190

Biswas HM, Patra PB, Boral MC (1981) Body fluid and hematologic changes in the toad exposed to 48 h of simulated high altitude. J Appl Physiol: Respir Environ Ex Physiol 51:794–797

Black CP, Tenney SM (1980) Oxygen transport during progressive hypoxia in high-altitude and sea-level waterfowl. Respir Physiol 39:217–239

Black CP, Tenney SM, Kroonenburg M van (1978) Oxygen transport during progressive hypoxia in bar-headed geese (*Anser indicus*) acclimated to sea level and 5600 m. In: Piiper J (ed) Respiratory function in birds, adult and embryonic. Springer, Berlin Heidelberg New York, pp 79–83

Bligh J, Johnson KG (1973) Glossary of terms for thermal physiology. J Appl Physiol 35:941–961

Block K (1962) Oxygen and biosynthetic patterns. Fed Proc 21:1058–1063

Blume FD, Pace N (1957) Effect of translocation to 3,800 meters altitude on glycolysis in mice. J Appl Physiol 23:75–79

Bock WJ (1980) The definition and recognition of biological adaptation. Am Zool 20:217–227

Boismare F (1979) Avoidance learning and mechanisms of protective effect of apomorphine under hypoxia. Acta Neurol Scand 60 (Suppl 72):160–161

Boismare F, Poncin M le, François J le, Hacpille L, Marchand JC (1977) Influence des catécholamines centrales sur l'aptitude fonctionnelle du rat en hypoxie. J Pharmacol (Paris) 8:287–296

Bolton TB (1979) Mechanisms of action of transmitters and other substances on smooth muscle. Physiol Rev 59:606–718

Bond AN (1960) An analysis of the response of salamander gills to changes in the oxygen concentration of the medium. Dev Biol 2:1–20

Booth JH (1979) The effect of oxygen supply, epinephrine and acetylcholine on the distribution of blood flow in trout gills. J Exp Biol 83:31–39

Bouverot P (1976) Rate of ventilatory acclimatization to altitude and strength of the O_2-chemoreflex drive. In: Duron B (ed) Respiratory centres and afferent systems. Edn INSERM (Paris) pp 213–219

Bouverot P (1978) Control of breathing in birds compared with mammals. Physiol Rev 58:604–655

Bouverot P, Bureau M (1975) Ventilatory acclimatizations and csf acid-base balance in carotid chemodenervated dogs at 3550 m. Pflügers Arch 361:17–23

Bouverot P, Fitzgerald RS (1969) Role of the arterial chemoreceptors in controlling lung volume in the dog. Respir Physiol 7:203–215

Bouverot P, Sébert P (1979) O_2-chemoreflex drive of ventilation in awake birds at rest. Respir Physiol 37:201–218

Bouverot P, Flandrois R, Puccinelli R, Dejours P (1965) Etude du rôle des chémorécepteurs artériels dans la régulation de la respiration pulmonaire chez le chien éveillé. Arch Int Pharmacodyn 157:253–271

Bouverot P, Candas V, Libert JP (1973) Role of the arterial chemoreceptors in ventilatory adaptation to hypoxia of awake dogs and rabbits. Respir Physiol 17:209–219

Bouverot P, Hildwein G, Oulhen P (1976) Ventilatory and circulatory O_2 convection at 4000 m in pigeon at neutral or cold temperature. Respir Physiol 28:371–385

Bouverot P, Douguet D, Sébert P (1979) Role of the arterial chemoreceptors in ventilatory and circulatory adjustments to hypoxia in awake Pekin ducks. J Comp Physiol 133:177–186

Bouverot P, Collin R, Favier R, Flandrois R, Sébert P (1981) Carotid chemoreceptor function in ventilatory and circulatory O_2 convection of exercising dogs at low and high altitude. Respir Physiol 43:147–167

Bowen WJ, Eads HJ (1949) Effects of 18000 feet simulated altitude on the myoglobin content of dogs. Am J Physiol 159:77–82

Bowes G, Townsend ER, Kozar LF, Bromley SM, Phillipson EA (1981) Effect of carotid body denervation on arousal response to hypoxia in sleeping dogs. J Appl Physiol: Respir Environ Ex Physiol 51:40–45

Brewer GJ, Oelshlegel FJ, Eaton WJ (1972) Biochemical, physiological, and genetic factors in the regulation of mammalian erythrocyte metabolism and DPG levels. In: Rorth M, Astrup P (eds) Oxygen affinity of hemoglobin and red cell acid base status. Munksgaard, Copenhagen, pp 539–551

Brody JS, Lahiri S, Simpser M, Motoyama EK, Velasquez T (1977) Lung elasticity and airway dynamics in Peruvian natives to high altitude. J Appl Physiol: Respir Environ Ex Physiol 42:245–251

Brody S (1945) Bioenergetics and growth. Reinhold Publ, New York. Repr 1968, Hafner Publ Comp, New York, p 1923

Brooks JG, Tenney SM (1968) Ventilatory response of llama to hypoxia at sea level and high altitude. Respir Physiol 5:269–278

Brouillette RT, Thach BT (1980) Control of genioglossus muscle inspiratory activity. J Appl Physiol: Respir Environ Ex Physiol 49:801–808

Brown RM, Engel J (1973) Evidence for catecholamines involvement in the suppression of locomotor activity due to hypoxia. J Pharm Pharmacol 25:815–819

Brusil PJ, Waggener TB, Kronauer RE, Gulesian P (1980) Methods for identifying respiratory oscillations disclose altitude effects. J Appl Physiol: Respir Environ Ex Physiol 48:545–556

Bui MV, Banchero N (1980) Effects of chronic exposure to cold or hypoxia on ventricular weights and ventricular myoglobin concentrations in guinea pigs during growth. Pflügers Arch 385:155–160

Buick FJ, Gledhill N, Froese AB, Spriet L, Meyers EC (1980) Effect of induced erythrocythemia on aerobic work capacity. J Appl Physiol: Respir Environ Ex Physiol 48:636–642

Bullard RW (1972) Vertebrates at altitudes. In: Yousef MK, Horvath SM, Bullard RW (eds) Physiological adaptations; desert and mountain. Academic Press, London New York, pp 209–225

Bullard RW, Kollias J (1966) Functional characteristics of two high altitude mammals. Fed Proc 25:1288–1292

Bunn HF (1980) Regulation of hemoglobin function in mammals. Am Zool 20:199–211

Bureau M, Bouverot P (1975) Blood and csf acid-base changes, and rate of ventilatory acclimatization of awake dogs to 3,550 m. Respir Physiol 24:203–216

Burggren W, Mwalukoma A (1983) Respiration during chronic hypoxia and hyperoxia in larval and adult bullfrogs (*Rana catesbeiana*). I. Morphological responses of lungs, skin and gills. J Exp Biol 105:191–203

Burri PH, Weibel ER (1971a) Morphometric estimation of pulmonary diffusion capacity. II. Effect of P_{O_2} on the growing lung. Respir Physiol 11:247–264

Burri PH, Weibel ER (1971b) Morphometric evaluation of changes in lung structure due to high altitude. In: Porter R, Knight J (eds) Ciba Found Symp. High altitude physiology: cardiac and respiratory aspects. Churchill Livingstone, London, pp 15–25

Burton RR, Besch EL, Smith AH (1968) Effect of chronic hypoxia on the pulmonary arterial blood pressure of the chicken. Am J Physiol 214:1438–1442

Buskirk ER (1978) Work capacity of high-altitude natives. In: Baker PT (ed) The biology of high altitude peoples. Cambridge Univ Press, Cambridge, pp 173–187

Cain SM, Chapler CK (1979) Oxygen extraction by canine hindlimb during hypoxic hypoxia. J Appl Physiol: Respir Environ Ex Physiol 46:1023–1028

Cain SM, Dunn JE (1966) Low doses of acetazolamide to acid accommodation of men to altitude. J Appl Physiol 21:1195–1200

Camba E, Montestruque S, Alvarez J (1980) Hypoxia and cold: influence on cellular oxygen-consuming systems in guinea pig liver. Life Sci 27:943–952

Capderou A, Polianski J, Mensch-Dechene J, Drouet L, Antezana G, Zelter M, Lockhart A (1977) Splanchnic blood flow, O_2 consumption, removal of lactate, and output of glucose in highlanders. J Appl Physiol: Respir Environ Ex Physiol 43:204–210

Capen RL, Wagner WW (1982) Intrapulmonary blood flow redistribution during hypoxia increases gas exchange surface area. J Appl Physiol: Respir Environ Ex Physiol 52:1575–1581

Carey C, Thompson EL, Vleck CM, James FC (1982) Avian reproduction over an altitudinal gradient: incubation period, hatchling mass, and embryonic oxygen consumption. Auk 99:710–718

Cascarano J, Ades IZ, O'Connor JD (1976) Hypoxia: a succinate-fumarate electron shuttle between peripheral cells and lungs. J Exp Zool 198:149–154

Cassin S, Gilbert RD, Bunnell CE, Johnson EM (1971) Capillary development during exposure to chronic hypoxia. Am J Physiol 220:448–451

Cerretelli P (1976) Limiting factors to oxygen transport on Mount Everest. J Appl Physiol 40:658–667

Cerretelli P (1980) Gas exchange at high altitude. In: West JB (ed) Pulmonary gas exchange, vol II. Academic Press, London New York, pp 97–147

Chabes A, Pareda J, Hyams L, Barrientos N, Perez J, Campos L, Monroe A, Mayorga A (1968) Comparative morphometry of the human placenta at high altitude and at sea level: the shape of the placenta. Obstet Gynecol 31:178–185

Chalmers JP, Korner PI, White SW (1967) The relative roles of the aortic and carotid sinus nerves in the rabbit in the control of respiration and circulation during arterial hypoxia and hypercapnia. J Physiol (London) 188:435–450

Chance B (1957) Cellular oxygen requirements. Fed Proc 16:671–680

Chance B (1965) Reaction of oxygen with the respiratory chain in cells and tissues. J Gen Physiol 49:163–188

Chance B (1977) Molecular basis of O_2 affinity for cytochrome oxidase. In: Jöbsis FF (ed) Oxygen and physiological function. Prof Inf Libr, Dallas, pp 14–25

Chance B, Cohen P, Jöbsis F, Schoener B (1962) Intracellular oxidation-reduction states in vivo. Science 137:499–508

Chance B, Oshino N, Sugano G, Mayevsky A (1973) Basic principles of tissue oxygenation determination from mitochondrial signals. In: Bucher HI, Brulve DF (eds) Oxygen transport to tissue. Plenum Press, New York London, pp 277–292

Chang AE, Detar R (1980) Oxygen and vascular smooth muscle contraction revisited. Am J Physiol 238 (Heart Circ Physiol 7):H716–H728

Cherniack NS, Edelman NH, Lahiri S (1970/71) Hypoxia and hypercapnia as respiratory stimulants and depressants. Respir Physiol 11:113–126

Chiodi H (1970/71) Comparative study of the blood gas transport in high altitude and sea level camelidae and goats. Respir Physiol 11:84–93

Cipriano LF, Pace N (1973) Glycolytic intermediates and adenosine phosphates in rat liver at high altitude (3,800 m). Am J Physiol 225:393–398

Clark DR, Smith P (1978) Capillary density and muscle fibre size in the hearts of rats subjected to simulated high altitude. Cardiovasc Res 12:578–584

Clegg EJ (1978) Fertility and early growth. In: Baker PT (ed) The biology of high-altitude peoples. Cambridge Univ Press, Cambridge, pp 65–115

Clench J, Ferrell RE, Schull WJ, Barton SA (1981) Hematocrit and hemoglobin, ATP and DPG concentrations in Andean man: the interaction of altitude and trace metals with glycolytic and hematologic parameters in man. In: Breer GJ (ed) The red cell. 5th Ann Arbor Conf Liss, New York, NY, pp 747–762
Clench J, Ferrell RE, Schull WJ (1982) Effect of chronic altitude hypoxia on hematologic and glycolytic parameters. Am J Physiol 242 (Regul Integr Comp Physiol 11):R447–R451
Cohen MM (1973) Biochemistry of cerebral anoxia, hypoxia, and ischemia. Monogr Neur Sci 1:1–49
Cohen PJ (1972) The metabolic function of oxygen and biochemical lesions of hypoxia. Anesthesiology 37:148–177
Colacino JM, Hector DH, Schmidt-Nielsen K (1977) Respiratory responses of ducks to simulated altitudes. Respir Physiol 29:265–281
Cole RP (1983) Skeletal muscle function in hypoxia: effect of alteration of intracellular myoglobin. Respir Physiol 53:1–14
Cole RP, Sukanek PC, Wittenberg JB, Wittenberg BA (1982) Mitochondrial function in the presence of myoglobin. J Appl Physiol: Respir Environ Ex Physiol 53:1116–1124
Cook SF, Alafi MH (1956) Role of the spleen in acclimatization to hypoxia. Am J Physiol 186:369–372
Costa LE, Taquini AC (1970) Effect of chronic hypoxia on myoglobin, cytochromes and ubiquinone levels in the rat. Acta Physiol Latinoam 20:103–109
Costa LE, Boveris A, Taquini AC (1982) Actividad respiratoria de mitocondrias cardiacas y hepaticas de rata en la hipoxia hipobarica cronica. Medicina (Buenos Aires) 42:259–264
Cotton EK, Hiestand M, Philbin GE, Simmons M (1980) Reevaluation of birth weights at high altitude. Am J Obstet Gynecol 138:220–222
Coulson RA, Herbert JD (1981) Relationship between metabolic rate and various physiological and biochemical parameters. A comparison of alligator, man and shrew. Comp Biochem Physiol 69A:1–13
Coulson RA, Hernandez T, Herbert JD (1977) Metabolic rate, enzyme kinetics in vivo. Comp Biochem Physiol 56A:251–262
Crowell JW, Smith EE (1967) Determinant of the optimal hematocrit. J Appl Physiol 22:501–504
Cruz JC, Hartley LH, Vogel JA (1975) Effect of altitude relocations upon AaD_{O_2} at rest and during exercise. J Appl Physiol 39:469–474
Cruz JC, Grover RF, Reeves JT, Maher JT, Cymerman A, Denniston JC (1976) Sustained venoconstriction in man supplemented with CO_2 at high altitude. J Appl Physiol 40:96–100
Cunningham EL, Brody JS, Jain BP (1974) Lung growth induced by hypoxia. J Appl Physiol 37:362–366
Cunningham WL, Becker EJ, Kreuzer F (1965) Catecholamines in plasma and urine at high altitude. J Appl Physiol 20:607–610
Curtin NA, Woledge RC (1978) Energy changes and muscular contraction. Physiol Rev 58:690–761
Daly M de Burgh, Scott MJ (1963) The cardiovascular responses to stimulation of the carotid body chemoreceptors in the dog. J Physiol (London) 165:179–197
Dam L Van (1938) On the utilization of oxygen and regulation of breathing in some aquatic animals. Ph D Thes, Volharding, Groningen, p 143
Davenport HW, Brewer G, Chambers AH, Goldschmidt S (1947) The respiratory responses to anoxemia of unanesthetized dogs with chronically denervated aortic and carotid chemoreceptors and their causes. Am J Physiol 148:406–416
Davis JN (1975) Adaptation of brain monoamine synthesis to hypoxia in the rat. J Appl Physiol 39:215–220
Davis JN (1976) Brain tyrosine hydroxylation: alteration of oxygen affinity in vivo by immobilization or electroshock in the rat. J Neurochem 27:211–215
Davis JN (1977) Synaptosomal tyrosine hydroxylation: affinity for oxygen. J Neurochem 28:1043–1050
Davis JN, Carlsson A (1973) Effect of hypoxia on tyrosine and tryptophan hydroxylation in unanesthetized rat brain. J Neurochem 20:913–915
Dawson A (1972) Regional lung function during early acclimatization to 3,100 m altitude. J Appl Physiol 33:218–223
Dawson A, Grover RF (1974) Regional lung function in natives and long-term residents at 3,100 m altitude. J Appl Physiol 36:294–298

DeGraff AC, Grover RF, Johnson RL, Hammond JW, Miller JM (1970) Diffusing capacity of the lung in Caucasians native to 3,100 m. J Appl Physiol 29:71–76

Dejours P (1957) Intérêt méthodologique de l'étude d'un organisme vivant à la phase initiale de rupture d'un équilibre physiologique. CR Acad Sci (Paris) 245:1946–1948

Dejours P (1962) Chemoreflexes in breathing. Physiol Rev 42:335–358

Dejours P (1976) Arterial chemoreceptors and ventilatory chemoreflexes in vertebrates. In: Duron B (ed) Respiratory centres and afferent systems. Edn INSERM, Paris, pp 205–211

Dejours P (1979) L'Everest sans oxygène: le problème respiratoire. J Physiol (Paris) 75:43A

Dejours P (1981) Principles of comparative respiratory physiology. Elsevier/North-Holland, Amsterdam, p 265

Dejours P (1982) Mount Everest and beyond: breathing air. In: Taylor CR, Johansen K, Bolis L (eds) A companion to animal physiology. Cambridge Univ Press, New York, NY, pp 17–30

Dejours P, Girard F, Labrousse Y, Teillac A (1959) Etude de la régulation de la ventilation de repos chez l'homme en haute altitude. Rev Franç Etudes Clin Biol 4:115–127

Dejours P, Kellogg RH, Pace N (1963) Regulation of respiration and heart rate response in exercise during altitude acclimatization. J Appl Physiol 18:10–18

Dejours P, Puccinelli R, Armand J, Dicharry M (1966) Breath-to-breath variations of pulmonary gas exchange in resting man. Respir Physiol 1:265–280

Dejours P, Garey WF, Rahn H (1970) Comparison of ventilatory and circulatory flow rates between animals in various physiological conditions. Respir Physiol 9:108–117

Dempsey JA, Forster HW (1982) Mediation of ventilatory adaptations. Physiol Rev 62:262–346

Dempsey JA, Reddan WG, Birnbaum ML, Forster HV, Thoden JS, Grover RF, Rankin J (1971) Effects of acute through life-long hypoxic exercise on exercise pulmonary gas exchange. Respir Physiol 13:62–89

Dempsey JA, Forster HV, doPico GA (1974) Ventilatory acclimatization to moderate hypoxemia in man: the role of spinal fluid [H^+]. J Clin Invest 53:1091–1100

Dempsey JA, Thomson JM, Forster HV, Cerny FC, Chosy LW (1975) HbO_2 dissociation in man during prolonged work in chronic hypoxia. J Appl Physiol 38:1022–1029

Dill DB, Evans DS (1970) Report barometric pressure. J Appl Physiol 29:914–916

Dill DB, Talbott JH, Consolazio WV (1937) Blood as a physicochemical system. XII Man at high altitudes. J Biol Chem 118:649–666

Dubach M (1981) Quantitative analysis of the respiratory system of the house sparrow, budgerigar and violet-eared hummingbird. Respir Physiol 46:43–60

Duffy TE, Nelson SR, Lowry OH (1972) Cerebral carbohydrate metabolism during acute hypoxia and recovery. J Neurochem 19:959–977

Duhm HF (1980) Regulation of hemoglobin function in mammals. Am Zool 20:199–211

Duhm J, Gerlach E (1971) On the mechanisms of the hypoxia-induced increase of 2,3-diphosphoglycerate in erythrocytes. Pflügers Arch 326:254–269

Duling RB (1978) Oxygen, metabolism, and microcirculatory control. In: Kaley G, Altura BM (eds) Microcirculation, vol II. Univ Park Press, Baltimore, pp 401–429

Duncker H-R (1972) Structure of avian lungs. Respir Physiol 14:44–63

Dunel-Erb S, Bailly Y, Laurent P (1982) Neuroepithelial cells in fish gill primary lamellae. J Appl Physiol: Respir Environ Ex Physiol 53:1342–1353

Durand J, Martineaud JP (1971) Resistance and capacitance vessels of the skin in permanent and temporary residents at high altitude. In: Porter R, Knight J (eds) High altitude physiology: cardiac and respiratory aspects. Churchill Livingstone, Edinburgh, pp 159–167

Durand J, Verpillat JP, Pradel M, Martineaud JP (1969) Influence of altitude on the cutaneous circulation of residents and newcomers. Fed Proc 28:1124–1128

Durand J, Marc-Vergnes JP, Coudert J, Blayo MC, Pocidalo JJ (1974) Cerebral blood flow, brain metabolism and csf acid-base balance in highlanders. In: Paintal AS, Gill-Kumar P (eds) Respiration adaptations, capillary exchange and reflex mechanisms. Proc Krogh Centenáry Symp, Srinagar, India. Univ Delhi, India, pp 27–37

Duve C de (1983) Microbodies in the living cell. Sci Am 248:52–62

Duve C de, Baudhuin P (1966) Peroxisomes (microbodies and related particles). Physiol Rev 46:323–357

Eaton JW, Skelton TD, Berger E (1974) Survival at extreme altitude: protective effect of increased hemoglobin oxygen affinity. Science 183:743–744

Eby SH, Banchero N (1976) Capillary density of skeletal muscle in Andean dogs. Proc Soc Exp Biol Med 151:795–798
Eclancher B (1975) Contrôle de la respiration chez les poissons téléostéens: réactions respiratoires à des changements rectangulaires de l'oxygénation du milieu. CR Acad Sci (Paris) 280:307–310
Else PL, Hulbert AJ (1981) Comparison of the "mammal machine" and the "reptile machine": energy production. Am J Physiol 240 (Regul Integr Comp Physiol 9):R3–R9
Elsner RW, Bolstad A, Forno C (1964) Maximum oxygen consumption of Peruvian Indians native to high altitude. In: Weihe WH (ed) The physiological effects of high altitude. Macmillan, New York, pp 217–223
Erasmus BD, Rahn H (1976) Effects of ambient pressures, He and SF6 on O_2 and CO_2 transport in the avian egg. Respir Physiol 27:53–64
Erslev AJ (1981) Erythroid adaptation to altitude. Blood Cells 7:495–508
Escobedo MA, Samaniego FC, Gonzales DV, Bernstein MH (1978) Respiration in pigeons at simulated high altitudes. Fed Proc 37:472
Ettinger RH, Staddon JER (1982) Decreased feeding associated with acute hypoxia in rats. Physiol Behav 29:455–458
Euler US von, Liljestrand G (1946) Observations on the pulmonary arterial blood pressure in the cat. Acta Physiol Scand 12:301–320
Farhi LE, Rahn H (1955) A theoretical analysis of the alveolar-arterial O_2 difference with special reference to the distribution effects. J Appl Physiol 7:699–703
Feigl EO (1983) Coronary physiology. Physiol Rev 63:1–205
Fencl V, Gabel RA, Wolfe D (1979) Composition of cerebral fluids in goats adapted to high altitude. J Appl Physiol: Respir Environ Ex Physiol 47:508–513
Fidelman ML, Seeholzer SH, Walsh KB, Moore RD (1982) Intracellular pH mediates action of insulin on glycolysis in frog skeletal muscle. Am J Physiol 242 (Cell Physiol 11):C87–C93
Fisher AB, Dodia C (1981) Lung as a model for evaluation of critical intracellular P_{O_2} and P_{CO}. Am J Physiol 241 (Endocrinol Metab 4):E47–E50
Fisher JW (1977) Kidney hormones, vol XI. Erythropoietin. Academic Press, London New York, p 601
Fishman AP (1976) Hypoxia on the pulmonary circulation. Circ Res 38:221–231
Folbergrova J, Nilsson B, Sakabe T, Siesjö BK (1981) The influence of hypoxia on the concentrations of cyclic nucleotides in the rat brain. J Neurochem 36:1670–1674
Folk GE (1966) Introduction to environmental physiology. Lea & Febiger, Philadelphia, p 308
Forster HV, Dempsey JA, Chosy LW (1975) Incomplete compensation of CSF $[H^+]$ in man during acclimatization to high altitude (4,300 m). J Appl Physiol 38:1067–1072
Forster HV, Bisgard GE, Rasmussen B, Orr JA, Buss DD, Manohar M (1976) Ventilatory control in peripheral chemoreceptor-denervated ponies during chronic hypoxemia. J Appl Physiol 41:878–885
Forster HV, Bisgard GE, Klein JP (1981) Effect of peripheral chemoreceptor denervation on acclimatization of goats during hypoxia. J Appl Physiol: Respir Environ Ex Physiol 50:392–398
Frayser R, Gray GW, Houston CS (1974) Control of the retinal circulation at altitude. J Appl Physiol 37:302–304
Freeman BM, Vince MA (1974) Development of the avian embryo. Chapman & Hall, London, p 362
Freminet A (1981) Carbohydrate and amino acid metabolism during acute hypoxia in rats: blood and heart metabolites. Comp Biochem Physiol 70B:427–433
Frisancho AR (1975) Functional adaptation to high altitude hypoxia. Science 187:313–319
Frisancho AR (1976) Growth and morphology at high altitude. In: Baker PT, Little MA (eds) Man in the Andes. Dowden, Hutchison & Ross, Strandsburg, Pn, pp 181–207
Gaehtgens P, Kreutz F, Albrecht KH (1979) Optimal hematocrit for canine skeletal muscle during rhythmic isotonic exercise. Eur J Appl Physiol 41:27–39
Gardiner M, Nilsson B, Rehncrona S, Siesjö BK (1981) Free fatty acids in the rat brain in moderate and severe hypoxia. J Neurochem 36:1500–1505
Garey WF, Rahn H (1970) Gas tensions in tissues of trout and carp exposed to diurnal changes in oxygen tension of the water. J Exp Biol 52:575–582
Garfinkel F, Fitzgerald RS (1978) The effect of hyperoxia, hypoxia, and hypercapnia on FRC and occlusion pressure in human subjects. Respir Physiol 33:241–250

Gatz RN, Piiper J (1979) Anaerobic energy metabolism during severe hypoxia in the lungless salamander *Desmognathus fuscus* (Plethodontidae). Respir Physiol 38:377–384
Gautier H, Bonora M (1980) Possible alterations in brain monoamine metabolism during hypoxia-induced tachypnea in cats. J Appl Physiol: Respir Environ Ex Physiol 49:769–777
Gautier H, Bonora M (1982) Effects of hypoxia and respiratory stimulants in conscious intact and carotid denervated cats. Bull Eur Physiopath Resp Clin Respir Physiol 18:565–582
Gautier H, Peslin R, Grassino A, Milic-Emili J, Hannhart B, Powell E, Miserocchi G, Bonora M, Fischer JT (1982) Mechanical properties of the lungs during acclimatization to altitude. J Appl Physiol: Respir Environ Ex Physiol 52:1407–1415
Gemmill CL, Reeves DL (1933) The effect of anoxemia in normal dogs before and after denervation of the carotid sinuses. Am J Physiol 105:487–495
Gerlach E, Duhm J (1972) 2,3-DPG metabolism of red cells: regulation and adaptive changes during hypoxia. In: Rorth M, Astrup P (eds) Oxygen affinity of hemoglobin and red cell acid base status. Munksgaard, Copenhagen, pp 552–569
Gibson GE, Blass JP (1976) Impaired synthesis of acetylcholine in brain accompanying mild hypoxia and hypoglycemia. J Neurochem 27:37–42
Gibson GE, Duffy TE (1981) Impaired synthesis of acetylcholine by mild hypoxic hypoxia or nitrous oxide. J Neurochem 36:28–33
Gibson GE, Jope R, Blass JP (1975) Decreased synthesis of acetylcholine accompanying impaired oxidation of pyruvic acid in rat brain minces. Biochem J 148:17–23
Gibson GE, Shimada M, Blass JP (1978) Alterations in acetylcholine synthesis and cyclic nucleotides in mild cerebral hypoxia. J Neurochem 31:757–760
Gibson GE, Peterson C, Sansone J (1981 a) Decreases in amino acid and acetylcholine metabolism during hypoxia. J Neurochem 37:192–201
Gibson GE, Peterson C, Sansone J (1981 b) Neurotransmitter and carbohydrate metabolism during aging and mild hypoxia. Neurobiol Aging 2:165–172
Gilbert RD, Cummings LA, Juchau MR, Longo LD (1979) Placental diffusing capacity and fetal development in exercising or hypoxic guinea pigs. J Appl Physiol: Respir Environ Ex Physiol 46:828–834
Gimenez M, Sanderson RJ, Reiss OK, Banchero N (1977) Effects of altitude on myoglobin and mitochondrial protein in canine skeletal muscle. Respiration 34:171–176
Gloster J, Heath D, Harris P (1972) The influence of diet on the effects of a reduced atmospheric pressure in the rat. Environ Physiol Biochem 2:117–124
Gold AJ, Costello LC (1974) Effects of altitude and semistarvation on heart mitochondrial function. Am J Physiol 227:1336–1339
Gold AJ, Johnson TF, Costello LC (1973) Effects of altitude stress on mitochondrial function. Am J Physiol 224:946–949
Goldstein M, Joh TH, Garvey TQ (1968) Kinetic studies of the enzymatic dopamine-beta-hydroxylation reaction. Biochemistry 7:2724–2730
Grandtner M, Turek Z, Kreuzer F (1974) Cardiac hypertrophy in the first generation of rats native to simulated high altitude. Pflügers Arch 350:241–248
Greaney GS, Place AR, Cashon RE, Smith G, Powers DA (1980) Time course of changes in enzyme activities and blood respiratory properties of killifish during long-term acclimation to hypoxia. Physiol Zool 53:136–144
Grollman A (1930) Physiological variations of the cardiac output of man. VII. The effect of high altitude on the cardiac output and its related functions: an account of experiments conducted on the summit of Pike's Peak, Colorado. Am J Physiol 93:19–40
Grover RF (1963) Basal oxygen uptake of man at high altitude. J Appl Physiol 18:909–912
Grover RF, Reeves JT, Will DH, Blount SG (1963) Pulmonary vasoconstriction in steers of high altitude. J Appl Physiol 18:567–574
Grover RF, Reeves JT, Grover EB, Leathers JE (1967) Muscular exercise in young men native to 3,100 m altitude. J Appl Physiol 22:555–564
Grover RF, Reeves JT, Maher JT, McCullough RE, Cruz JC, Denniston JC, Cymerman A (1976 a) Maintained stroke volume but impaired arterial oxygenation in man at high altitude with supplemental CO_2. Circ Res 38:391–396
Grover RF, Lufschanowski R, Alexander JK (1976 b) Alterations in the coronary circulation of man following ascent to 3,100 m altitude. J Appl Physiol 41:832–838

Grover RF, Tucker A, Reeves JT (1982) Hypobaria: an etiologic factor in acute mountain sickness? In: Loeppky JA, Riedesel ML (eds) Oxygen transport to human tissues. Elsevier/North-Holland, New York, pp 223–230

Grubb BR (1981) Blood flow and oxygen consumption in avian skeletal muscle during hypoxia. J Appl Physiol: Respir Environ Ex Physiol 50:450–455

Grubb BR, Colacino JM, Schmidt-Nielsen K (1978) Cerebral blood flow in birds: effect of hypoxia. Am J Physiol 234 (Heart Circ Physiol 3):H230–H234

Grunewald WA, Sowa W (1977) Capillary structures and O_2 supply to tissue. An analysis with a digital diffusion model as applied to the skeletal muscle. Rev Physiol Biochem Pharmacol 77:149–209

Guazzi M, Freis ED (1969) Sino-aortic reflexes and arterial pH, P_{O_2}, and P_{CO_2} in wakefulness and sleep. Am J Physiol 217:1623–1627

Guleria JS, Pande JN, Sethi PK, Roy SB (1971) Pulmonary diffusing capacity at high altitude. J Appl Physiol 31:536–543

Gurtner GH, Burns B (1972) Possible facilitated transport of oxygen across the placenta. Nature (London) 240:473–475

Guz A, Noble MIM, Widdicombe JG, Trenchard D, Mushin WW (1966) Peripheral chemoreceptor in man. Respir Physiol 1:38–40

Haab P, Held DR, Ernst H, Farhi LE (1969) Ventilation-perfusion relationships during high-altitude adaptation. J Appl Physiol 26:77–81

Haas JD (1981) Human adaptability approach to nutritional assessment: a Bolivian example. Fed Proc 40:2577–2582

Hackett PH, Rennie D, Grover RF, Reeves JT (1981) Acute mountain sickness and the edemas of high altitude: a common pathogenesis. Respir Physiol 46:383–390

Haddy FJ, Scott JB (1968) Metabolically linked vasoactive chemicals in local regulation of blood flow. Physiol Rev 48:688–707

Haldane JS, Kellas AM, Kennaway EL (1919) Experiments on acclimatization to reduced atmospheric pressure. J Physiol (London) 53:181–206

Hall FG (1966) Minimal utilizable oxygen and the oxygen dissociation curve of blood of rodents. J Appl Physiol 21:375–378

Hall FG, Dill DB, Guzman Barron ES (1936) Comparative physiology in high altitudes. J Cell Comp Physiol 8:301–313

Hannon JP (1981) Nutrition at high altitude. In: Horvath SM, Yousef MK (eds) Environmental physiology: aging, heat, and altitude. Elsevier/North-Holland, New York, pp 309–327

Hannon JP, Vogel JA (1977) Oxygen transport during early altitude acclimatization: a perspective study. Eur J Appl Physiol 36:285–297

Harper AM, McCulloch J (1977) Dopaminergic influence upon the cerebral circulation in the anesthetized baboon. J Physiol (London) 265:24P–25P

Harris P, Castillo Y, Gibson K, Heath D, Arias Stella J (1970) Succinic and lactic dehydrogenase activity in myocardial homogenates from animal at high and low altitude. J Mol Cell Cardiol 1:189–193

Hartley LH, Vogel JA, Cruz JC (1974) Reduction of maximal exercise heart rate at altitude and its reversal with atropine. J Appl Physiol 36:362–365

Hatcher JD, Chiu LK, Jennings DB (1978) Anemia as a stimulus to aortic and carotid chemoreceptors in the cat. J Appl Physiol: Respir Environ Ex Physiol 44:696–702

Hawkins RA, Biebuyck JF (1979) Ketone bodies are selectively used by individual brain regions. Science 205:325–327

Hayashi M, Nagasaka T (1981) Enhanced heat production in physically restrained rats in hypoxia. J Appl Physiol: Respir Environ Ex Physiol 51:1601–1606

Heath D (1982) The lung at high altitude. In: Bonsignore G, Cumming G (eds) The lung in its environment. Plenum Press, New York London, pp 447–457

Heath D, Williams DR (1981) Man at high altitude. The pathophysiology of acclimatization and adaptation. Churchill Livingstone, Edinburgh, p 347

Hebbel RP, Eaton JW, Kronenberg RS, Zanjani ED, Moore LG, Berger EM (1978) Human llamas. Adaptation to altitude in subjects with high hemoglobin oxygen affinity. J Clin Invest 62:593–600

Hedner T (1978) Central monoamine metabolism and neonatal oxygen deprivation. Acta Physiol Scand Suppl 460

Hedner T, Lundborg P (1979) Regional changes in monoamine synthesis in the developing rat brain during hypoxia. Acta Physiol Scand 106:139–143

Hedner T, Lundborg P, Engel J (1977) Effect of hypoxia on monoamine synthesis in brains of developing rats. Biol Neonate 31:122–126

Helleger A, Metcalfe J, Huckabee WE, Prystowsky H, Meschia G, Barron DH (1961) Alveolar P_{CO_2} and P_{O_2} in pregnant and nonpregnant women at high altitude. Am J Obstet Gynecol 82:241–245

Hellung-Larsen P, Jensen MA, Sorensen SC (1973) Changes in the lactate dehydrogenase isoenzyme pattern of human brains following lifelong exposure to hypoxia. Acta Physiol Scand 87:15A–16A

Hesse R (1947) Ecological animal geography. Transl Allee WC, Schmidt KP, 3rd edn. Wiley & Sons, New York, p 597

Heusner AA (1972) Criteria for standard metabolism. Proc Int Symp Environ Physiol (Bioenerget). FASEB, pp 15–21

Heusner AA (1982a) Energy metabolism and body size. I: Is the 0.75 mass exponent of Kleiber's equation a statistical artifact? Respir Physiol 48:1–12

Heusner AA (1982b) Energy metabolism and body size. II: Dimensional analysis and energetic nonsimilarity. Respir Physiol 48:13–25

Heusner AA (1983) Body size, energy metabolism and the lungs. J Appl Physiol: Respir Environ Ex Physiol 54:867–873

Heymans C, Neil E (1958) Reflexogenic areas of the cardiovascular system. Churchill, London, p 271

Hill JR (1959) The oxygen consumption of new-born and adult mammals. Its dependence on the oxygen tension in the inspired air and on the environmental temperature. J Physiol (London) 149:346–373

Hillyard SD (1981) Respiratory and cardiovascular adaptations of amphibians and reptiles to altitude. In: Horvath SM, Yousef MK (eds) Environmental physiology: aging, heat and altitude. Elsevier/North-Holland, New York, pp 362–377

Hinkle PC, McCarty RE (1978) How cells make ATP? Sci Am 238:104–123

Hochachka PW (1972) Comparative intermediary metabolism. In: Prosser CL (ed) 3rd edn. Saunders, Philadelphia, Pa, pp 212–278

Hochachka PW (1980) Living without oxygen. Harvard Univ Press, Cambridge, Ma, p 181

Hochachka PW, Somero GN (1973) Strategies of biochemical adaptation. Saunders, Philadelphia, Pa, p 358

Hochachka PW, Owen TG, Allen JF, Whittow GC (1975) Multiple end products of anaerobiosis in diving vertebrates. Comp Biochem Physiol 50B:17–22

Hochachka PW, Stanley C, Merkt J, Sumar-Kalinowski J (1983) Metabolic meaning of elevated levels of oxidative enzymes in high altitude adapted animals: an interpretive hypothesis. Respir Physiol 52:303–313

Holeton GF (1980) Oxygen as an environmental factor of fishes. In: Ali MA (ed) Environmental physiology of fishes. Plenum Press, New York London, pp 7–32

Holeton GF, Randall DJ (1967) The effect of hypoxia upon the partial pressure of gases in the blood and water afferent and efferent to the gills of rainbow trout. J Exp Biol 46:317–327

Holland RAB, Forster RE (1966) The effect of size of red cells on the kinetics of their oxygen uptake. J Gen Physiol 49:727–742

Holloszy JO (1975) Adaptation of skeletal muscle to endurance exercise. Med Sci Sports 7:155–164

Hoon RS, Balasubramanian V, Mathew OP, Tiwari SC, Sharma SC, Chadha KS (1977) Effect of high-altitude exposure for 10 days on stroke volume and cardiac output. J Appl Physiol: Respir Environ Ex Physiol 42:722–727

Hornbein TF (ed) (1981) Regulation of breathing (2 vols). Dekker, New York, p 1436

Horstman D, Weiskopf R, Jackson RE (1980) Work capacity during 3-wk sojourn at 4,300 m: effects of relative polycythemia. J Appl Physiol: Respir Environ Ex Physiol 49:311–318

Horvath SM, Jackson RL (1982) Significance of P_{50} under stress. In: Loeppky JA, Riedesel ML (eds) Oxygen transport to human tissues. Elsevier/North-Holland, New York, NY, pp 305–315

Howe A, Pack RJ, Wise JCM (1981) Arterial chemoreceptor-like activity in the abdominal vagus of the rat. J Physiol (London) 320:309–318

Hudlická O (1982) Growth of capillaries in skeletal and cardiac muscle. Circ Res 50:451–461

Hugelin A, Bonvallet M, Dell P (1959) Activation réticulaire et corticale d'origine chémoceptive au cours de l'hypoxie. Electroenceph Clin Neurophysiol 11:325–340

Hughes GM, Morgan M (1973) The structure of fish gills in relation to their respiratory function. Biol Rev 48:419–475

Hughes MJ, Light KE, Redington T (1983) Alterations in CNS amine levels by acclimatization to hypobaric hypoxia. Brain Res Bull 11:255–258

Hultgren HN, Grover RF (1968) Circulatory adaptation to high altitude. Annu Rev Med 19:119–152

Hultgren HN, Marticorena EA (1978) High altitude pulmonary edema. Epidemiologie observations in Peru. Chest 74:372–376

Hurtado A (1932) Studies at high altitude. Blood observations of the Indian natives of the Peruvian Andes. Am J Physiol 100:487–505

Hurtado A (1964) Animals in high altitude: resident man. In: Dill DB, Adolph EF, Vilber CG (eds) Handbook of physiology, Sect 4. Adaptation to the environment. Am Physiol Soc, Washington, DC, pp 843–860

Hurtado A, Rotta A, Merino C, Pons J (1937) Studies of myoglobin at high altitudes. Am J Med Sci 194:708–713

Hurtado A, Merino C, Delgado E (1945) Influence of anoxemia on the hemopoietic activity. Arch Int Med 75:284–323

Hutchison VH, Haines HB, Engbretson G (1976) Aquatic life at high altitude: respiratory adaptations in the lake Titicaca frog, *Telmatobius culeus*. Respir Physiol 27:115–129

Imbert G, Hildwein G, Dejours P (1976) Breath-to-breath variations of alveolar P_{O_2} and P_{CO_2} at barometric pressures of 490, 745 and 1500 Torr in resting awake dogs. Respir Physiol 28:207–216

Ingermann RL, Stock MK, Metcalfe J, Shih T-B (1983) Effect of ambient oxygen on organic phosphate concentrations in erythrocytes of the chick embryo. Respir Physiol 51:141–152

International Civil Aviation Organization (1964) Manual of the ICAO standard atmosphere. US Gov Print Off, Washington, DC

Ishii K, Honda K, Ishii K (1966) The function of the carotid labyrinth in the toad. Tohoku J Exp Med 88:103–106

Issekutz B, Shaw WAS, Issekutz TB (1975) Effect of lactate on FFA and glycerol turnover in resting and exercising dogs. J Appl Physiol 39:349–353

Itazawa Y, Takeda T (1978) Gas exchange in the carp gills in normoxic and hypoxic conditions. Respir Physiol 35:263–269

Jackson DC (1973) Ventilatory response to hypoxia in turtles at various temperatures. Respir Physiol 18:178–187

Jaeger JJ, Sylvester JT, Cymerman A, Berberich JJ, Denniston JC, Maher JT (1979) Evidence for increased intrathoracic fluid volumes in man at high altitude. J Appl Physiol: Respir Environ Ex Physiol 47:670–676

Jaeger N (1979) Carnets de solitude. Denoel, Paris, p 236

Jameson EW, Heusner AA, Arbogast R (1977) Oxygen consumption of *Sceloporus occidentalis* from three different elevations. Comp Biochem Physiol 56A:73–79

Jöbsis FF (1964) Basic processes in cellular respiration. In: Fenn WO, Rahn H (eds) Handbook of physiology, sect 3. Respiration, vol I. Am Physiol Soc, Washington, DC, pp 63–124

Jöbsis FF (1972) Oxidative metabolism at low P_{O_2}. Fed Proc 31:1404–1413

Jöbsis FF (1979) Oxidative metabolic effects of cerebral hypoxia. In: Fahn S, Davis JN, Rowland LP (eds) Advances in neurology, vol XXVI. Cerebral hypoxia and its consequences. Raven Press, New York, pp 299–318

Jöbsis FF, Keizer JH, LaManna JC, Rosenthal M (1977) Reflectance spectrophotometry of cytochrome aa_3 in vivo. J Appl Physiol: Respir Environ Ex Physiol 43:858–872

Johansen K (1979) Cardiovascular support of metabolic functions in vertebrates. In: Wood SC, Lenfant C (eds) Evolution of respiratory processes; a comparative approach. Dekker, New York, pp 107–192

Johansen K (1982) Respiratory gas exchange of vertebrate gills. In: Houlihan DF, Rankin JC, Shuttleworth TJ (eds) Gills. Cambridge Univ Press, New York, pp 99–128

Jones DP, Kennedy FG (1982) Intracellular oxygen supply during hypoxia. Am J Physiol 243 (Cell Physiol 12):C247–C253

Jones DR, Purves MJ (1970) The effect of carotid body denervation upon the respiratory response to hypoxia and hypercapnia in the duck. J Physiol (London) 211:295–309

Jones NL, Robertson DG, Kane JW, Hart RA (1972) Effect of hypoxia on free fatty acid metabolism during exercise. J Appl Physiol 33:733–738

Jones RM, LaRochelle FT, Tenney SM (1981) Role of arginine vasopressin on fluid and electrolyte balance in rats exposed to high altitude. Am J Physiol 240 (Regul Integr Comp Physiol 9):R182–R186

Kaaja R, Are K (1982) Myocardial LDH isoenzyme patterns in rats exposed to cold and/or hypobaric hypoxia. Acta Med Scand (Suppl 668):136–142

Kaley G (1978) Microcirculatory-endocrine interactions. Role of prostaglandins. In: Kaley G, Altura BM (eds) Microcirculation, vol II. Univ Park Press, Baltimore, Md, pp 503–529

Kalsner S (1977) The effect of hypoxia on prostaglandin output and on tone in isolated coronary arteries. Can J Physiol Pharmacol 55:882–887

Katz IR (1980) Oxygen affinity of tyrosine and tryptophane hydroxylase in synaptosomes. J Neurochem 35:760–763

Katz IR (1982) Oxygen dependence of dopamine-β-hydroxylase activity and lactate metabolism in synaptosomes from rat brain. Brain Res 231:399–409

Kaufman S, Fisher DP (1974) Pterin-requiring aromatic amino acid hydroxylases. In: Hayaishi O (ed) Molecular mechanisms of oxygen activation. Academic Press, London New York, pp 285–369

Kawashiro T, Nusse W, Scheid P (1975) Determination of diffusivity of oxygen and carbon dioxide in respiring tissue: results in rat skeletal muscle. Pflügers Arch 359:231–251

Kearney MS (1973) Ultrastructural changes in the heart at high altitude. Pathol Microbiol 39:258–265

Kellogg RH (1968) Altitude acclimatization, a historical introduction emphasizing the regulation of breathing. Physiologist 11:37–57

Kellogg RH (1977) Oxygen and carbon dioxide in the regulation of respiration. Fed Proc 36:1658–1663

Kellogg RH, Mines AH (1975) Acute hypoxia fails to affect FRC in man. Physiologist 18:275

Keynes RJ, Smith GW, Slater JDH, Brown MM, Brown SE, Payne NN, Jowett TP, Monge CC (1982) Renin and aldosterone at high altitude in man. J Endocrinol 92:131–140

Keys A, Hall FG, Barron ES (1936) The position of the oxygen dissociation curve of human blood at high altitude. Am J Physiol 115:292–307

Kinnula VL (1976) Mitochondrial cytochrome concentrations in rat heart and liver as a consequence of different hypoxic periods. Acta Physiol Scand 96:417–421

Kinnula VL, Hassinen IE (1977) Effects of hypoxia on mitochondrial mass and cytochrome concentrations in rat heart and liver during postnatal development. Acta Physiol Scand 99:462–466

Kinnula VL, Hassinen IE (1978) Effect of chronic hypoxia on hepatic triacylglycerol concentration and mitochondrial fatty acid oxidizing capacity in liver and heart. Acta Physiol Scand 102:64–73

Kinnula VL, Hassinen IE (1981) Effects of hypoxia and fasting on the cytochrome concentration in intestinal epithelial villous cell mitochondria. Acta Physiol Scand 112:387–393

Kleiber M (1961) The fire of life, an introduction to animal energetics. Wiley & Sons, New York, p 454

Knoblauch A, Sybert A, Brennan NJ, Sylvester JT, Gurtner GH (1981) Effect of hypoxia and CO on a cytochrome P-450-mediated reaction in rabbit lungs. J Appl Physiol: Respir Environ Ex Physiol 51:1635–1642

Knox WE, Auerbach VH, Lin ECC (1956) Enzymatic and metabolic adaptations in animals. Physiol Rev 36:164–255

Kock LL de (1959) The carotid body system of the higher vertebrates. Acta Anat 37:265–279

Koepchen HP, Klussendorf D, Borchert J, Lessmann DW, Dinter A, Frank C, Sommer D (1976) Type of respiratory neuronal activity pattern in reflex control of ventilation. In: Paintal AS, Gill-Kumar P (eds) Respiratory adaptations, capillary exchange and reflex mechanisms. Vallabhbhai Patel Chest Inst, Delhi, pp 291–311

Koford CB (1956) The vicuña and the puna. Ecol Monogr 26:153–219

Kontos HA, Levasseur JE, Richardson DW, Mauck HP, Patterson JL (1967) Comparative circulatory response to systemic hypoxia in man and in unanesthetized dog. J Appl Physiol 23:381–386

Koob GF, Annau Z (1973) Effect of hypoxia on hypothalamic mechanisms. Am J Physiol 224:1403–1408

Korner PI (1971) Integrative neural cardiovascular control. Physiol Rev 51:312–367

Korner PI, White SW (1960) Circulatory control in hypoxia by the sympathetic nerves and adrenal medulla. J Physiol (London) 184:272–290

Kramer K, Luft UC (1951) Mobilization of red cells and oxygen from the spleen in severe hypoxia. Am J Physiol 165:215–228

Krasney JA, Koehler RC (1980) Neural control of the circulation during hypoxia. In: Hughes MJ, Barnes CD (eds) Research topics in physiology, V, 2: Neural control of circulation. Academic Press, London New York, pp 123–147

Kreuzer F (1966) Transport of O_2 and CO_2 at altitude. In: Margaria R (ed) Exercise at altitude. Excerpta Med Found, Amsterdam, pp 149–158

Kreuzer F (1970) Facilitated diffusion of oxygen and its possible significance; a review. Respir Physiol 9:1–30

Kreuzer F, Lookeren Campagne P van (1965) Resting pulmonary diffusing capacity for CO and O_2 at high altitude. J Appl Physiol 20:519–524

Kreuzer F, Tenney SM, Andresen DC, Schreiner BF, Nye RE, Mithoefer JC, Valtin H, Naitove A (1960) Alveolar-arterial oxygen gradient in the dog at altitude. J Appl Physiol 15:796–800

Kreuzer F, Tenney SM, Mithoefer JC, Remmers J (1964) Alveolar-arterial oxygen gradient in Andean natives at high altitude. J Appl Physiol 19:13–16

Krogh A (1922) The anatomy and physiology of capillaries. Yale Univ Press, New Haven, p 276

Kronenberg RS, Safar P, Lee J, Wright F, Noble W, Wahrenbrock E, Hickley R, Nemoto E, Severinghaus J (1971) Pulmonary artery pressure and alveolar gas exchange in man during acclimatization to 12,470 ft. J Clin Invest 50:827–837

Krüger H, Arias-Stella J (1970) The placenta and the newborn infant at high altitudes. Am J Obstet Gynecol 106:586–591

Kuramoto K, Matsushita S, Matsuda T, Mifune J, Sakai M, Iwasaki T, Shinagawa T, Moroki N, Murakami M (1980) Effect of hematocrit and viscosity on coronary circulation and myocardial oxygen utilization. Jpn Circ J 44:443–448

Kuschinsky W, Wahl M (1978) Local chemical and neurogenic regulation of cerebral vascular resistance. Physiol Rev 58:656–690

Lahiri S (1968) Alveolar gas pressures in man with life-time hypoxia. Respir Physiol 4:373–386

Lahiri S (1972) Unattenuated ventilatory hypoxic drive in ovine and bovine species native to high altitude. J Appl Physiol 32:95–102

Lahiri S (1975) Blood oxygen affinity and alveolar ventilation in relation to body weight in mammals. Am J Physiol 229:529–536

Lahiri S (1976) Depressant effect of acute and chronic hypoxia on ventilation. In: Paintal AS (ed) Morphology and mechanisms of chemoreceptors. Vallabhbhai Patel Chest Inst, Delhi, pp 138–145

Lahiri S (1980) Role of arterial O_2 flow in peripheral chemoreceptor excitation. Fed Proc 39:2648–2652

Lahiri S, Edelman NH (1969) Peripheral chemoreflexes in the regulation of breathing of high altitude natives. Respir Physiol 6:375–385

Lahiri S, Gelfand R (1981) Mechanisms of acute ventilatory responses. In: Hornbein T (ed) Lung biology in health and disease, part 2, vol XVII. Dekker, New York, pp 773–844

Lahiri S, Milledge JS, Sorensen SC (1972) Ventilation in man during exercise at high altitude. J Appl Physiol 32:766–769

Lahiri S, Brody JS, Motoyama EK, Velasquez TM (1978) Regulation of breathing in newborn at high altitude. J Appl Physiol: Respir Environ Ex Physiol 44:673–678

Lahiri S, Edelman NH, Cherniack NS, Fishman AP (1981) Role of carotid chemoreflex in respiratory acclimatization to hypoxemia in goat and sheep. Respir Physiol 46:367–382

Lahiri S, Maret K, Sherpa MG (1983) Dependence of high altitude sleep apnea on ventilatory sensitivity to hypoxia. Respir Physiol 52:281–301

Lauweryns JM, Cokelaere M (1973) Hypoxia-sensitive neuroepithelial bodies. Intrapulmonary secretory neuroreceptors, modulated by the CNS. Z Zellforsch 145:521–540

Lauweryns JM, Cokelaere M, Theunynck P (1972) Neuroepithelial bodies in the respiratory mucosa of various mammals. Z Zellforsch 135:569–592

Lechner AJ (1976) Respiratory adaptations in burrowing pocket gophers from sea level and high altitude. J Appl Physiol 41:168–173

Lechner AJ (1977) Metabolic performance during hypoxia in native and acclimated pocket gophers. J Appl Physiol: Respir Environ Ex Physiol 43:965–970

Lechner AJ, Banchero N (1980) Lung morphometry in guinea pigs acclimated to hypoxia during growth. Respir Physiol 42:155–169

Lechner AJ, Salvato VL, Banchero N (1980) Hematology and red cell morphology in guinea pigs acclimated to chronic hypoxia during growth. Comp Biochem Physiol 67A:239–244
Lechner AJ, Salvato VL, Banchero N (1981) The hematological response to hypoxia in growing guinea pigs is blunted during concomitant cold stress. Comp Biochem Physiol 70A:321–327
Lechner AJ, Grimes MJ, Aquin L, Banchero N (1982) Adaptative lung growth during chronic cold plus hypoxia is age-dependent. J Exp Zool 219:285–291
Lefrançois R, Gautier H, Pasquis P (1968) Ventilatory oxygen drive in acute and chronic hypoxia. Respir Physiol 4:217–228
Lehninger AL (1970) Biochemistry: the molecular basis of cell structure and function. Worth Publ, New York, p 833
Leitner LM, Pagès B, Puccinelli R, Dejours P (1965) Etude simultanée de la ventilation et des décharges des chémorécepteurs du glomus carotidien chez le chat. I. Au cours d'inhalations brèves d'oxygène pur. Arch Int Pharmacodyn 154:421–426
Lenfant C (1973) High altitude adaptation in mammals. Am Zool 13:447–456
Lenfant C, Sullivan K (1971) Adaptation to high altitude. New Engl J Med 284:1298–1309
Lenfant C, Torrance J, English E, Finch CA, Reynafarje C, Ramos J, Faura J (1968) Effect of altitude on oxygen binding by hemoglobin and on organic phosphate levels. J Clin Invest 47:2652–2656
Lenfant C, Torrance JD, Reynafarje C (1971) Shift of the O_2-Hb dissociation curve at altitude: mechanism and effect. J Appl Physiol 30:625–631
Levasseur JE, Kontos HA, Richardson DW, Patterson JL (1976) Circulatory effects of prolonged hypoxia before and during antihistamine. J Appl Physiol 40:549–558
Lockhart A, Saiag B (1981) Altitude and the human pulmonary circulation. Clin Sci 60:599–605
Lomholt JP, Johansen K (1979) Hypoxia acclimation in carp. How it affects O_2 uptake, ventilation, and O_2 extraction from water. Physiol Zool 52:38–49
Longo LD, Bartels H (eds) (1972) Respiratory gas exchange and blood flow in the placenta. US Department of Health, Education, and Welfare. DHEW Publ No (NIH) 73–361, pp 570
Longo LD, Hill EP, Power GG (1972) Theoretical analysis of factors affecting placental O_2 transfer. Am J Physiol 222:730–739
Lozano R, Monge CC (1965) Renal function in high-altitude natives and in natives with chronic mountain sickness. J Appl Physiol 20:1026–1027
Lübbers DW (1977) Quantitative measurement and description of oxygen supply to the tissue. In: Jöbsis FF (ed) Oxygen and physiological function. Prof Inf Libr, Dallas, pp 254–276
Luft UC (1964) Aviation physiology. The effects of altitude. In: Fenn WO, Rahn H (eds) Handbook of physiology, sect 3. Respiration, vol II. Am Physiol Soc, Washington, DC, pp 1099–1145
Lutz PL (1980) On the oxygen affinity of bird blood. Am Zool 20:187–198
Lutz PL, Schmidt-Nielsen K (1977) Effect of simulated altitude on blood gas transport in the pigeon. Respir Physiol 30:383–388
Mager M, Blatt WF, Natale PJ, Blatteis CM (1968) Effect of high altitude on lactic dehydrogenase isozymes of neonatal and adult rats. Am J Physiol 215:8–13
Mani MS (1962) Introduction to high altitude entomology. Methuen, London, p 302
Mani MS (1974) High altitude insects. In: Fundamentals of high altitude biology. Oxford IBH Publ Co, New Delhi
Manohar M, Parks CM, Busch MA, Tranquilli WJ, Bisgard GE, McPherron TA, Theodorakis MC (1982) Regional myocardial blood flow and coronary vascular reserve in unanesthetized young calves exposed to a simulated altitude of 3500 m for 8–10 weeks. Circ Res 50:714–726
Maren TH (1967) Carbonic anhydrase: chemistry, physiology, and inhibition. Physiol Rev 47:595–781
Markelonis G, Garbus J (1975) Alterations of intracellular oxidative metabolism as stimuli evoking prostaglandin biosynthesis. Prostaglandins 10:1087–1106
Mazess RB (1975) Human adaptation to high altitude. In: Damon A (ed) Physiological anthropology. Oxford Univ Press, New York, pp 167–209
McCutcheon IE, Metcalfe J, Metzenberg AB, Ettinger T (1982) Organ growth in hyperoxic and hypoxic chick embryos. Respir Physiol 50:153–163
McDonald DM (1981) Peripheral chemoreceptors. Structure-function relationships of the carotid body. In: Hornbein TF (ed) Regulation of breathing, part 1. Dekker, New York, pp 105–319
McFadden DM, Houston CH, Sutton JR, Powles ACP, Gray GW, Roberts RS (1981) High-altitude retinopathy. J Am Med Assoc 245:581–586

McFarland RA, Evans JN (1939) Alterations in dark adaptation under reduced oxygen tensions. Am J Physiol 127:37–50

McGrath JJ (1971) Acclimation response of pigeons to simulated high altitude. J Appl Physiol 31:274–276

McGrath RL, Weil JV (1978) Adverse effects of normovolemic polycythemia and hypoxia on hemodynamics in the dog. Circ Res 43:793–798

McMurtry IF, Rounds S, Stanbrook HS (1982) Studies of the mechanism of hypoxic pulmonary vasoconstriction. Adv Shock Res 8:21–33

Meerson FZ (1975) Role of synthesis of nucleic acids and protein in adaptation to the external environment. Physiol Rev 55:79–123

Mela L, Goodwin CW, Miller LD (1976) In vivo control of mitochondrial enzyme concentrations and activity by oxygen. Am J Physiol 231:1811–1816

Mensen de Silva E, Cazorla A (1973) Lactate, α-GP, and Krebs cycle in sea-level and high-altitude native guinea pigs. Am J Physiol 224:669–672

Merino CF (1950) Studies on blood formation and destruction in the polycythemia of high altitude. Blood 5:1–31

Meschia G, Prystowsky H, Hellegers A, Huckabee W, Metcalfe J, Barron DH (1960) Observations on the oxygen supply to the fetal llama. Q J Exp Physiol 45:284–291

Messner R (1978) Everest. Expedition zum Endpunkt. BLV Verlagsges, München

Metcalfe J, Dhindsa DS (1970) A comparison of mechanism of oxygen transport among several mammalian species. In: Brewer GJ (ed) Red cell metabolism and function. Plenum Press, New York London, pp 229–241

Metcalfe J, Meschia G, Hellegers A, Prystowsky H, Huckabee W, Barron DH (1962) Observations on the placental exchange of the respiratory gases in pregnant ewes at high altitude. Q J Exp Physiol 47:74–92

Metcalfe J, Bartels H, Moll W (1967) Gas exchange in the pregnant uterus. Physiol Rev 47:782–838

Metcalfe J, Bissonnette JM, Bowles RE, Matsumoto JA, Dunham SJ (1979) Hen's eggs with retarded gas exchange. I. Chorioallantoic capillary growth. Respir Physiol 36:97–101

Miles DS, Bransford DR, Horvath SM (1981) Hypoxia effects on plasma volume shifts at rest, work, and recovery in supine posture. J Appl Physiol: Respir Environ Ex Physiol 51:148–153

Milic-Emili J, Grunstein MM (1976) Drive and time components of ventilation. Chest (Suppl) 70:131–133S

Miller AT, Hale DM (1968) Organ lactic dehydrogenase in altitude-acclimatized rats. J Appl Physiol 25:725–728

Miller AT, Hale DM (1970) Increased vascularity of brain, heart, and skeletal muscle of polycythemic rats. Am J Physiol 219:702–704

Miller MJ, Tenney SM (1975) Hypoxia-induced tachypnea in carotid-deafferented cats. Respir Physiol 23:31–39

Mitchell RA, Sinha AK, McDonald DM (1972) Chemoreceptive properties of regenerated endings of the carotid sinus nerve. Brain Res 43:681–685

Mithoefer JC, Remmers JE, Zubieta G, Mithoefer MC (1972) Pulmonary gas exchange in Andean natives at high altitude. Respir Physiol 15:182–189

Monge C Sr, Monge C Jr (1968) Adaptation to high altitude. In: Hafez ESE (ed) Adaptation of domestic animals. Lea & Febiger, Philadelphia, Pa, pp 194–201

Monge C, Whittembury J (1976) High altitude adaptations in the whole animal. In: Bligh J, Cloudsley-Thompson JL, Macdonald AG (eds) Environmental physiology of animals. Blackwell, Oxford, pp 289–308

Monge MC (1943) Chronic mountain sickness. Physiol Rev 23:166–184

Monge CC, Lozano R, Marchena C, Whittembury J, Torres C (1969) Kidney function in the high-altitude native. Fed Proc 28:1199–1203

Moore LG, Jahnigen D, Rounds SS, Reeves JT, Grover RF (1982a) Maternal hyperventilation helps preserve arterial oxygenation during high-altitude pregnancy. J Appl Physiol: Respir Environ Ex Physiol 52:690–694

Moore LG, Roundo SS, Jahnigen D, Grover RF, Reeves JT (1982b) Infant birth weight is related to maternal arterial oxygenation at high altitude. J Appl Physiol: Respir Environ Ex Physiol 52:695–699

Moret PR (1971) Coronary blood flow and myocardial metabolism in man at high altitude. In: Porter R, Knight J (eds) High altitude physiology: cardiac and respiratory aspects. Churchill Livingstone, Edinburgh, pp 131–144

Moret PR, Bopp P, Righetti A, Bloch A, Suter P (1981) Entraînement physique et haute altitude. Schweiz Med Wochenschr 111:1693–1696

Morpurgo G, Arese P, Bosia A, Pescarmona GP, Luzzana M, Modiano G, Ranjit SK (1976) Sherpas living permanently at high altitude: a new pattern of adaptation. Proc Natl Acad Sci USA 73:747–751

Morrison P (1964) Wild animals at high altitudes. Symp Zool Soc (London) 13:49–55

Mulligan E, Lahiri S, Storey BT (1981) Carotid body O_2 chemoreception and mitochondrial oxidative phosphorylation. J Appl Physiol: Respir Environ Ex Physiol 51:438–446

Napier J (1972) Bigfoot. The Yeti and Sasquatch in myth and reality. Cape, London, p 240

Nelson BD, Highman B, Altland PD (1967) Oxidative phosphorylation during altitude acclimation in rats. Am J Physiol 213:1414–1418

Nesarajah MS, Matalon S, Krasney JA, Farhi LE (1983) Cardiac output and regional oxygen transport in the acutely hypoxic conscious sheep. Respir Physiol 53:161–172

Neubauer JA, Santiago TV, Edelman NH (1981) Hypoxic arousal in intact and carotid chemodenervated sleeping cats. J Appl Physiol: Respir Environ Ex Physiol 51:1294–1299

Neville JR (1977a) Altered haem-haem interaction and tissue-oxygen supply: a theoretical analysis. Br J Haematol 35:387–395

Neville JR (1977b) Theoretical analysis of altitude tolerance and hemoglobin function. Aviat Space Environ Med 48:409–412

Newsholme EA, Crabtree B (1981) Flux-generating and regulatory steps in metabolic control. Trends Biochem Sci 121:53–56

Nice P van, Black CP, Tenney SM (1980) A comparative study of ventilatory responses to hypoxia with reference to hemoglobin O_2-affinity in llama, cat, rat, duck and goose. Comp Biochem Physiol 66A:347–350

Niedenzu C, Grasedyck K, Voelkel NF, Bittmann S, Lindner J (1981) Proliferation of lung cells in chronically hypoxic rats. Int Arc Occup Environ Health 48:185–193

Nienhuis AW, Benz EJ (1977) Regulation of hemoglobin synthesis during the development of the red cell. New Engl J Med 297:1318–1328

Nikolov R, Nikolova M, Miyares C, Milanova D (1982) Antihypoxic effect of prostacyclin. Meth Find Exp Clin Pharmacol 4:211–219

Nishino T, Lahiri S (1981) Effects of dopamine on chemoreflexes in breathing. J Appl Physiol: Respir Environ Ex Physiol 50:892–897

Norberg K, Siesjö BK (1975) Cerebral metabolism in hypoxic hypoxia. I. Pattern of activation of glycolysis: a reevaluation. Brain Res 86:31–44

Nylander E, Lund N, Wranne B (1983) Effect of increased blood oxygen affinity on skeletal muscle surface oxygen pressure fields. J Appl Physiol: Respir Environ Ex Physiol 54:99–104

Olson EB, Dempsey JA (1978) Rat as a model for humanlike ventilatory adaptation to chronic hypoxia. J Appl Physiol: Respir Environ Ex Physiol 44:763–769

Opitz E (1951) Increased vascularization of the tissue due to acclimatization to high altitude and its significance for the oxygen transport. Exp Med Surg 9:389–403

Orr JA, Bisgard GE, Forster HV, Buss DD, Dempsey JA, Will JA (1975) Cerebrospinal fluid alkalosis during high altitude sojourn in unanesthetized ponies. Respir Physiol 25:23–37

Ou LC (1974) Hepatic and renal gluconeogenesis in rats acclimatized to high altitude. J Appl Physiol 36:303–307

Ou LC, Smith RP (1983) Probable strain differences of rats in susceptibilities and cardiopulmonary responses to chronic hypoxia. Respir Physiol 53:367–377

Ou LC, Tenney SM (1970) Properties of mitochondria from hearts of cattle acclimatized to high altitude. Respir Physiol 8:151–159

Ou LC, Kim D, Layton WM, Smith RP (1980) Splenic erythropoiesis in polycythemic response of the rat to high-altitude exposure. J Appl Physiol: Respir Environ Ex Physiol 48:857–861

Pace N (1974) Respiration at high altitude. Fed Proc 33:2126–2132

Packard GC, Sotherland PR, Packard MJ (1977) Adaptive reduction in permeability of avian eggshells to water vapour at high altitudes. Nature (London) 266:255–256

Paganelli CV, Ar A, Rahn H, Wangensteen OD (1975) Diffusion in the gas phase: the effects of ambient pressure and gas composition. Respir Physiol 25:247–258

Paintal AS (ed) (1976) Morphology and mechanisms of chemoreceptors. Vallabhbhai Patel Chest Inst, Univ Delhi, p 357
Paulo LG, Fink GD, Roh BL, Fisher JW (1973) Influence of carotid body ablation on erythropoietin production in rabbits. Am J Physiol 224:442–444
Pawson IG, Jest C (1978) The high-altitude areas of the world and their cultures. In: Baker PT (ed) The biology of high-altitude peoples. Cambridge Univ Press, Cambridge, pp 17–45
Pearson OP (1951) Mammals in the highlands of Southern Peru. Bull Mus Comp Zool 106:117–173
Pearson OP, Bradford DF (1976) Thermoregulation of lizards and toads at high altitudes in Peru. Copeia 1976:155–170
Penney DG (1974) Lactate dehydrogenase subunit and activity changes in hypertrophied heart of the hypoxically exposed rat. Biochim Biophys Acta 358:21–24
Penney DG, Thomas M (1975) Hematological alterations and response to acute hypobaric stress. J Appl Physiol 39:1034–1037
Petschow D, Wurdinger I, Baumann R, Duhm J, Braunitzer G, Bauer C (1977) Causes of high blood O_2 affinity of animals living at high altitude. J Appl Physiol: Respir Environ Ex Physiol 42:139–143
Pettersson K, Johansen K (1982) Hypoxic vasoconstriction and the effects of adrenaline on gas exchange efficiency in fish gills. J Exp Biol 97:263–272
Picon-Reategui E (1966) Insulin, epinephrin, and glucagon on the metabolism of carbohydrates at high altitude. Fed Proc 25:1233–1238
Piiper J, Scheid P (1971) Maximum gas transfer efficacy of models for fish gills, avian lungs and mammalian lungs. Respir Physiol 14:115–124
Piiper J, Scheid P (1975) Gas transport efficacy of gills, lungs and skin: theory and experimental data. Respir Physiol 23:209–221
Piiper J, Scheid P (1977) Comparative physiology of respiration: functional analysis of gas exchange organs in vertebrates. In: Widdicombe JG (ed) International review of physiology, vol XIV. Univ Park Press, Baltimore, pp 219–253
Piiper J, Scheid P (1980) Blood-gas equilibration in lungs. In: West JB (ed) Pulmonary gas exchange, vol I. Ventilation, blood flow, and diffusion. Academic Press, London New York, pp 131–171
Piiper J, Scheid P (1981) Model for capillary-alveolar equilibration with special reference to O_2 uptake in hypoxia. Respir Physiol 46:193–208
Piiper J, Scheid P (1982) Physical principles of respiratory gas exchange in fish gills. In: Houlihan DF, Rankin JC, Shuttleworth TJ (eds) Gills. Cambridge Univ Press, New York, pp 45–61
Piiper J, Cerretelli P, Cuttica F, Mangili F (1966) Energy metabolism and circulation in dogs exercising in hypoxia. J Appl Physiol 21:1143–1149
Piiper J, Dejours P, Haab P, Rahn H (1971) Concepts and basic equations in gas exchange physiology. Respir Physiol 13:292–304
Piiper J, Tazawa H, Ar A, Rahn H (1980) Analysis of chorioallantoic gas exchange in the chick embryo. Respir Physiol 39:273–284
Pinder A, Burggren W (1983) Respiration during chronic hypoxia and hyperoxia in larval and adult bullfrogs (*Rana catesbeiana*). II. Changes in respiratory properties of whole blood. J Exp Biol 105:205–213
Podgorski GT, Longmuir IS, Knopp JA, Benson DM (1981) Use of an encapsulated fluorescent probe to measure intracellular P_{O_2}. J Cell Physiol 107:329–334
Poel WE (1949) Effect of anoxic anoxia on myoglobin concentration in striated muscle. Am J Physiol 156:44–51
Popel AS (1982) Oxygen diffusive shunts under conditions of heterogeneous oxygen delivery. J Theor Biol 96:533–541
Pough FH (1980) Blood oxygen transport and delivery in reptiles. Am Zool 20:173–185
Powell FL (1982) Diffusion in avian lungs. Fed Proc 41:2131–2133
Powell FL, Wagner PD (1982) Ventilation-perfusion in avian lungs. Respir Physiol 48:233–241
Power GC, Longo LD, Wagner HN, Kuhl DE, Forester RE (1967) Uneven distribution of maternal and fetal placenta blood flow, as demonstrated using macro-aggregates, and its response to hypoxia. J Clin Invest 46:2053–2063
Prampero PE di (1981) Energetics of muscular exercise. Rev Physiol Biochem Pharmacol 89:143–222
Prampero PE di, Mognoni P, Veicsteinas A (1982) The effects of hypoxia on maximal anaerobic alactic power in man. In: Brendel W, Zink RA (eds) High altitude physiology and medicine. Springer, Berlin Heidelberg New York, pp 88–93

Prosser CL (1964) Perspectives of adaptation: theoretical aspects. In: Dill DB, Adolph EF, Vilber CG (eds) Handbook of physiology, sect 4. Adaptation to environment. Am Physiol Soc, Washington, DC, pp 11–25

Prosser CL (1973) Comparative animal physiology, 3rd edn. Saunders, Philadelphia, Pa, p 966

Pugh LGCE (1957) Resting ventilation and alveolar air on Mount Everest: with remarks on the relation of barometric pressure to altitude in mountains. J Physiol (London) 135:590–610

Pugh LGCE (1964) Animals in high altitudes: man above 5,000 meters-mountain exploration. In: Dill DB (ed) Handbook of physiology, sect 4. Adaptation to the environment. Am Physiol Soc, Washington, DC, pp 861–868

Pugh LGCE, Gill MB, Lahiri S, Milledge JS, Ward MP, West JB (1964) Muscular exercise at great altitudes. J Appl Physiol 19:431–440

Purves MJ (ed) (1975) The peripheral arterial chemoreceptors. Cambridge Univ Press, Cambridge, p 492

Racker E (1980) From Pasteur to Mitchell: a hundred years of bioenergetics. Fed Proc 39:210–215

Rahn H (1949) A concept of mean alveolar air and the ventilation-blood flow relationships during pulmonary gas exchange. Am J Physiol 158:21–30

Rahn H (1966) Aquatic gas exchange: Theory. Respir Physiol 1:1–12

Rahn H (1977) Adaptation of the avian embryo to altitude. In: Paintal AS, Gill-Kumar P (eds) Respiratory adaptations, capillary exchange and reflex mechanisms. Proc Krogh Centenary Symp, Srinagar, India. Univ Delhi, India, pp 94–105

Rahn H (1983) Altitude adaptation: organisms without lungs. In: Hypoxia, exercise, and altitude. Liss, New York, pp 345–363

Rahn H, Ar A (1974) The avian egg: incubation time and water loss. Condor 76:147–152

Rahn H, Farhi LE (1964) Ventilation, perfusion, and gas exchange. The $\dot{V}_A/\dot{Q}$ concept. In: Fenn WO, Rahn H (eds) Handbook of physiology, sect 3, respiration, vol I. Am Physiol Soc, Washington, DC, pp 735–766

Rahn H, Fenn WO (1955) A graphical analysis of the respiratory gas exchange. Am Physiol Soc, Washington, DC, p 38

Rahn H, Otis AB (1949) Man's respiratory response during and after acclimatization to high altitude. Am J Physiol 157:445–462

Rahn H, Wangensteen OD, Crowley GJ (1973) Tissue O_2 and CO_2 tensions of trout in high altitude lakes. Trans Am Fish Soc 102:132–134

Rahn H, Carey C, Balmas K, Bhatia B, Paganelli C (1977) Reduction of pore area of the avian eggshell as an adaptation to altitude. Proc Natl Acad Sci USA 74:3095–3098

Rahn H, Ar A, Paganelli CV (1979) How bird eggs breathe. Sci Am 240:38–47

Rakusan K, Turek Z, Kreuzer F (1981) Myocardial capillaries in guinea pigs native to high altitude (Junin, Peru, 4,105 m). Pflügers Arch 391:22–24

Randall DJ (1970) Gas exchange in fish. In: Hoar WS, Randall DJ (eds) Fish physiology, vol IV. Academic Press, London New York, pp 253–292

Randall DJ (1982) The control of respiration and circulation in fish during exercise and hypoxia. J Exp Biol 100:275–288

Randall DJ, Jones DR (1973) The effect of deafferentation of the pseudobranch on the respiratory response to hypoxia and hyperoxia in the trout (*Salmo gairdneri*). Respir Physiol 17:291–301

Reed DJ, Kellogg RH (1960) Effect of sleep on hypoxic stimulation of breathing at sea level and altitude. J Appl Physiol 15:1130–1134

Reed RD, Pace N (1980) Energy status and oxidation-reduction status in rat liver at high altitude (3.8 km). Aviat Space Environ Med 51:595–602

Reeves JT, Halpin J, Cohn JE, Daoud F (1969) Increased alveolar-arterial oxygen difference during simulated high-altitude exposure. J Appl Physiol 27:658–661

Reeves RB (1977) The interaction of body temperature and acid-base balance in ectothermic vertebrates. Annu Rev Physiol 39:559–586

Reeves RB, Rahn H (1979) Patterns in vertebrate acid-base regulation. In: Wood SC, Lenfant C (eds) Evolution of respiratory processes: a comparative approach. Dekker, New York, pp 225–252

Rehncrona S, Siesjö BK, Westerberg E (1978) Adenosine and cyclic AMP in cerebral cortex of rats in hypoxia, status epilepticus and hypercapnia. Acta Physiol Scand 104:453–463

Reich JG, Sel'Kov EE (1981) Energy metabolism of the cell. A theoretical treatise. Academic Press, London New York, p 345

Reid L (1967) The embryology of the lung. In: Reuck AVS de, Porter R (eds) Development of the lung. Ciba Found Symp. Churchill, London, pp 109–124
Reid RC, Sherwood TK (1966) The properties of gases and liquids, 2nd edn. McGraw-Hill, New York, p 646
Reite M, Jackson D, Cahoon R, Weil JV (1975) Sleep physiology at high altitude. Electroencephalogr Clin Neurophysiol 38:463–471
Remmers JE (1981) Control of breathing during sleep. In: Hornbein TF (ed) Regulation of breathing, part 2. Dekker, New York, pp 1197–1249
Remmers JE, Mithoefer JC (1969) The carbon monoxide diffusing capacity in permanent residents at high altitudes. Respir Physiol 6:233–244
Reynafarje B (1962) Myoglobin content and enzymatic activity of muscle and altitude adaptation. J Appl Physiol 17:301–305
Reynafarje BD (1971) Mecanismos moleculares de la adaptacion a la hipoxia de las grandes alturas. Arch Inst Biol Andina 4:1–14
Reynafarje C (1964) Hematologic changes during rest and physical activity in man at high altitude. In: Weihe WH (ed) The physiological effects of high altitude. Pergamon Press, Oxford New York, pp 73–85
Reynafarje C (1966) Iron metabolism during and after altitude exposure in man and in adapted animals (camelids). Fed Proc 25:1240–1242
Reynafarje C, Faura J, Villavicencio D, Curaca A, Reynafarje B, Oyola L, Contreras L, Vallenas E, Faura A (1975) Oxygen transport of hemoglobin in high-altitude animal (Camelidae). J Appl Physiol 38:806–810
Riar SS, Shankar Bhat K, Sen Gupta J (1982) Physiological responses of mules on prolonged exposure to high altitude (3650 m). Int J Biometeor 26:129–136
Richards DW (1960) Homeostasis: its dislocations and perturbations. Perspect Biol Med 3:238–251
Richardson TQ, Guyton AC (1959) Effect of polycythemia and anemia on cardiac output and other circulatory factors. Am J Physiol 197:1167–1170
Ristori MT, Laurent P (1977) Action de l'hypoxie sur le système vasculaire branchial de la tête perfusée de truite. CR Soc Biol 171:809–813
Robin ED (1980) Of men and mitochondria: coping with hypoxic dysoxia. Am Rev Respir Disease 122:517–531
Robinson DS (1980) Changes in monoamine oxidase and monoamines with human development and aging. Fed Proc 3A:103–107
Rodeau JL, Malan A (1979) A two-compartment model of blood acid-base state at constant or variable temperature. Respir Physiol 37:5–30
Roos A, Boron WF (1981) Intracellular pH. Physiol Rev 61:296–434
Rosenmann M, Morrison P (1974) Physiological responses to hypoxia in the tundra vole. Am J Physiol 227:734–739
Rosenmann M, Morrison PR (1975) Metabolic response of highland and lowland rodents to simulated high altitudes and cold. Comp Biochem Physiol 51A:523–530
Rosenthal M, LaManna JC, Jöbsis FF, Levasseur JE, Kontos HA, Patterson JL (1976) Effect of respiratory gases on cytochrome a in intact cerebral cortex: is there a critical P_{O_2}? Brain Res 108:143–154
Ross JM, Fairchild HM, Weldy J, Guyton AC (1962) Autoregulation of blood flow by oxygen lack. Am J Physiol 202:21–24
Roth JA (1979) Effect of drugs on inhibition of oxidized and reduced form of MAO. In: Singer TP (ed) Monoamine oxidase: structure, function, and altered functions. Academic Press, London New York, pp 153–168
Roughton FJW, Forster RE (1957) Relative importance of diffusion and chemical reaction rates in determining rate of exchange of gases in the human lung, with special reference to true diffusing capacity of pulmonary membrane and volume of blood in the lung capillaries. J Appl Physiol 11:290–302
Ruben JA, Battalia DE (1979) Aerobic and anaerobic metabolism during activity in small rodents. J Exp Zool 208:73–76
Rubin RP, Laychock SG (1978) Prostaglandins and calcium membrane interactions in secretory glands. Ann NY Acad Sci 307:377–389

Rubio R, Berne RM, Bockman EL, Curnish RR (1975) Relationship between adenosine concentration and oxygen supply in rat brain. Am J Physiol 228:1896–1902
Ruiz AV, Bisgard GE, Will JA (1973) Hemodynamic responses to hypoxia and hyperoxia in calves at sea level and altitude. Pflügers Arch 344:275–286
Russell JA, Crook L (1968) Comparison of metabolic responses of rats to hypoxic stress produced by two methods. Am J Physiol 214:1113–1116
Saligaut C, Moore N, Leclerc JL, Boismare F (1981) Dopaminergic agonists and conditioned avoidance response in normoxia or hypoxic rats. Aviat Space Environ Med 52:19–23
Saltin B, Grover RF, Blomqvist CG, Hartley LH, Johnson RL (1968) Maximal oxygen uptake and cardiac output after two weeks at 4300 m. J Appl Physiol 25:400–409
Saltin B, Nygaard E, Rasmussen B (1980) Skeletal muscle adaptation in man following prolonged exposure to high altitude. Acta Physiol Scand 109:31A
Samaja M, Veicsteinas A, Cerretelli P (1979) Oxygen affinity of blood in altitude Sherpas. J Appl Physiol: Respir Environ Ex Physiol 47:337–341
Sanborn T, Gavin W, Berkowitz S, Perille T, Lesch M (1979) Augmented conversion of aspartate and glutamate to succinate during anoxia in rabbit heart. Am J Physiol 237 (Heart Circ Physiol 6):H535–H541
Scheid P (1979) Mechanisms of gas exchange in bird lungs. Rev Physiol Biochem Pharmacol 86:137–186
Scheid P (1982) Respiration and control of breathing. In: Farner DS, King JR, Parkes KC (eds) Avian biology, vol XI. Academic Press, London New York, pp 405–453
Scheid P (1984, to be published) Significance of lung structure for performance at high altitude. Ornithol Congr, Moscow
Scheid P, Piiper J (1980) Intrapulmonary gas mixing and stratification. In: West JB (ed) Pulmonary gas exchange, vol I. Ventilation, blood flow, and diffusion. Academic Press, London New York, pp 87–130
Schmidt-Nielsen K, Pennycuik P (1961) Capillary density in mammals in relation to body size and oxygen consumption. Am J Physiol 200:746–750
Schneider EC (1921) Physiological effects of altitude. Physiol Rev 1:631–659
Sears TA, Berger AJ, Phillipson EA (1982) Reciprocal tonic activation of inspiratory and expiratory motoneurones by chemical drives. Nature (London) 299:728–730
Serra I, Alberghina M, Viola M, Giuffrida AM (1981) Effect of hypoxia on nucleic acid and protein synthesis in different brain regions. Neurochem Res 6:595–605
Severinghaus JW (1972) Hypoxic respiratory drive and its loss during chronic hypoxia. Clin Physiol 2:57–79
Severinghaus JW, Mitchell RA, Richardson BW, Singer MM (1963) Respiratory control at high altitude suggesting active transport regulation of CSF pH. J Appl Physiol 18:1155–1166
Severinghaus JW, Bainton CR, Carcelen A (1966 a) Respiratory insensitivity to hypoxia in chronically hypoxic man. Respir Physiol 1:308–334
Severinghaus JW, Chiodi H, Eger EI, Brandstater B, Hornbein TF (1966 b) Cerebral blood flow in man at high altitude. Circ Res 29:274–282
Shapiro W, Wasserman AJ, Baker JP, Patterson JL (1970) Cerebrovascular responses to acute hypocapnic and eucapnic hypoxia in normal man. J Clin Invest 49:2362–2368
Shappell SD, Lenfant CJM (1975) Physiological role of the oxyhemoglobin dissociation curve. In: Surgenor DM (ed) The red blood cell, 2nd edn, vol II. Academic Press, London New York, pp 841–871
Shelton G (1970) The regulation of breathing. In: Hoar WS, Randall DJ (eds) Fish physiology, vol IV. Academic Press, London New York, pp 293–359
Shertzer HG, Cascarano J (1972) Mitochondrial alterations in heart, liver, and kidney of altitude-acclimated rats. Am J Physiol 223:632–636
Siesjö BK (1977) Brain metabolism in relation to oxygen supply. In: Jöbsis FF (ed) Oxygen and physiological function. Prof Inf Libr, Dallas, pp 459–479
Siesjö BK (1978) Brain energy metabolism. Wiley & Sons, Chichester, p 607
Sillau AH, Banchero N (1977) Effects of hypoxia on capillary density and fiber composition in rat skeletal muscle. Pflügers Arch 370:227–232
Sillau AH, Aquin L, Bui MV, Banchero N (1980a) Chronic hypoxia does not affect guinea pig skeletal muscle capillarity. Pflügers Arch 386:39–45

Sillau AH, Cueva S, Morales P (1980b) Pulmonary arterial hypertension in male and female chickens at 3300 m. Pflügers Arch 386:269–275
Silver IA (1981) Oxygen tension in the clinical situation. In: Gilbert DL (ed) Oxygen and living processes. Springer, Berlin Heidelberg New York, pp 358–367
Sime F, Penaloza D, Ruiz L, Gonzales N, Covarrubias E, Postigo R (1974) Hypoxemia, pulmonary hypertension, and low cardiac output in newcomers at low altitude. J Appl Physiol 36:561–565
Sinha NDP, Dejours P (1979) Effect of wide range of water oxygenation on breathing and blood acid-base status of crayfish *Astacus leptodactylus*. Indian J Exp Biol 17:888–891
Smialek M, Hamberger A (1970) The effect of moderate hypoxia and ischemia on cytochrome oxidase activity and protein synthesis in brain mitochondria. Brain Res 17:369–371
Smith AH (1973) Avian physiology. In: Kellogg RH (ed) 25 years of high altitude research: White Mountain Research station. Univ Cal, Berkeley, pp 19–22
Smith AH, Burton RR, Besch EL (1969) Development of the chick embryo at high altitude. Fed Proc 28:1092–1098
Smith PG, Mills E (1980) Restoration of reflex ventilatory response to hypoxia after removal of carotid bodies in the cat. Neuroscience 5:573–580
Smith RH, Guilbeau EJ, Reneau DD (1977) The oxygen field within a discrete volume of cerebral cortex. Microvasc Res 13:233–240
Smith RP, Kruszyna R, Ou L-C (1979) Hemoglobinemia in mice exposed to high altitude. Pflügers Arch 380:65–70
Snyder GK, Weathers WW (1977) Activity and oxygen consumption during hypoxic exposure in high altitude and lowland sceloporine lizards. J Comp Physiol 117:291–301
Snyder GK, Black CP, Birchard GF (1981) Physiologic adaptation to hypoxia in the avian embryo. Physiologist 24:135
Snyder GK, Black CP, Birchard GF (1982a) Development and metabolism during hypoxia in embryos of high-altitude *Anser indicus* versus sea-level *Branta canadensis* geese. Physiol Zool 55:113–123
Snyder GK, Black CP, Birchard GF, Lucich R (1982b) Respiratory properties of blood from embryos of highland vs lowland geese. J Appl Physiol: Respir Environ Ex Physiol 53:1432–1438
Snyder LRG (1982) 2,3-diphosphoglycerate in high- and low-altitude populations of the deer mouse. Respir Physiol 48:107–123
Snyder LRG, Born S, Lechner AJ (1982) Blood oxygen affinity in high- and low-altitude populations of the deer mouse. Respir Physiol 48:89–105
Soivio A, Nikinmaa M, Westman K (1980) The blood oxygen binding properties of hypoxic *Salmo gairdneri*. J Comp Physiol 136:83–87
Sørensen SC, Lassen NA, Severinghaus JW, Coudert J, Zamora MP (1974) Cerebral glucose metabolism and cerebral blood flow in high-altitude residents. J Appl Physiol 37:305–310
Squires RW, Buskirk ER (1982) Aerobic capacity during acute exposure to simulated altitude, 914 to 2286 meters. Med Sci Sports Ex 14:36–40
Stahl WR (1967) Scaling of respiratory variables in mammals. J Appl Physiol 22:453–460
Stainsby WN, Otis AB (1964) Blood flow, blood oxygen tension, oxygen uptake, and oxygen transport in skeletal muscle. Am J Physiol 206:858–866
Staub NC, Bishop JM, Forster RE (1962) Importance of diffusion and chemical reaction rated in O_2 uptake in lung. J Appl Physiol 17:21–27
Staudte HW, Pette D (1972) Correlations between enzymes of energy-supplying metabolism as a basic pattern of organization in muscle. Comp Biochem Physiol 41B:533–540
Steinbrook RA, Donovan JC, Gabel RA, Leith DE, Fencl V (1983) Acclimatization to high altitude in goats with ablated carotid bodies. J Appl Physiol: Respir Environ Ex Physiol 55:16–21
Stenberg J, Ekblom B, Messin R (1966) Hemodynamic response to work at simulated altitude, 4000 m. J Appl Physiol 21:1589–1594
Stere AJ, Anthony A (1977) Myocardial Feulgen-DNA levels and capillary vascularization in hypoxia-exposed rats. J Appl Physiol 42:501–507
Stickney JC, Liere EJ van (1953) Acclimatization to low oxygen tension. Physiol Rev 33:13–34
St. John WM, Wang SC (1977) Response of medullary respiratory neurons to hypercapnia and isocapnic hypoxia. J Appl Physiol 43:812–821
Strickland EH, Ackerman E, Anthony A (1962) Respiration and phosphorylation in liver and heart mitochondria from altitude-exposed rats. J Appl Physiol 17:535–538

Strohl KP (1982) Periodic breathing and sleep hypoxemia. In: Sutton JR, Jones NL, Houston CS (eds) Hypoxia: man at altitude. Thieme-Stratton, New York, pp 102–105
Strother GK, Ackerman E, Anthony A, Strickland E (1959) Effects of altitude acclimatization on rat myoglobin. Effect of viscosity and acclimatization on myoglobin reaction rates. Am J Physiol 196:517–519
Sugano T, Oshino N, Chance B (1974) Mitochondrial functions under hypoxic conditions. The steady state of cytochrome c reduction and of energy metabolism. Biochim Biophys Acta 347:340–358
Surks MI, Chinn KSK, Matoush LO (1966) Alterations in body composition in man after acute exposure to high altitude. J Appl Physiol 21:1741–1746
Susheela L, Ramasarma T (1973) Nature of the activation of succinate dehydrogenase by various effectors and in hypobaria and hypoxia. Biochim Biophys Acta 292:50–63
Sutton JR, Gray GW, Houston CS, Powles ACP (1980) Effects of duration at altitude and acetazolamide on ventilation and oxygenation during sleep. Sleep 3:455–464
Sutton JR, Jones NL, Pugh LGCE (1983) Exercise at altitude. Annu Rev Physiol 45:427–437
Swan LW (1961) The ecology of the high Himalayas. Sci Am 205:67–78
Sylvester JT, Cymerman A, Gurtner G, Hottenstein O, Cote M, Wolfe D (1981) Components of alveolar-arterial O_2 gradient during rest and exercise at sea level and high altitude. J Appl Physiol: Respir Environ Ex Physiol 50:1129–1139
Tappan DV, Reynafarje B (1957) Tissue pigment manifestations of adaptation to high altitudes. Am J Physiol 190:99–103
Tappan DV, Reynafarge BD, Potter VR, Hurtado A (1957) Alterations in enzymes and metabolites resulting from adaptation to low oxygen tensions. Am J Physiol 190:93–98
Teisseire BP, Soulard CD, Hericault RA, Leclerc LF, Laver MB (1979) Effect of chronic changes in hemoglobin-O_2 affinity in rats. J Appl Physiol: Respir Environ Ex Physiol 46:816–822
Tenney SM (1974) A theoretical analysis of the relationship between venous blood and mean tissue oxygen pressures. Respir Physiol 20:283–296
Tenney SM, Ou LC (1970) Physiological evidence for increased tissue capillarity in rats acclimatized to high altitude. Respir Physiol 8:137–150
Tenney SM, Ou LC (1977a) Ventilatory response of decorticate and decerebrate cats to hypoxia and CO_2. Respir Physiol 29:81–92
Tenney SM, Ou LC (1977b) Hypoxic ventilatory response of cats at high altitude: an interpretation of "blunting." Respir Physiol 30:185–199
Tenney SM, Remmers JE (1966) Alveolar dimensions in the lungs of animals raised at high altitude. J Appl Physiol 21:1328–1330
Tenney SM, John WM St (1980) Is there localized cerebral cortical influence on hypoxic ventilatory response? Respir Physiol 41:227–232
Tenney SM, Rahn H, Stroud RC, Mithoefer JC (1953) Adaptation to high altitude: changes in lung volumes during the first seven days at Mt. Evans, Colorado. J Appl Physiol 5:607–613
Tenney SM, Scotto P, Ou LC, Bartlett D, Remmers JE (1971) Suprapontine influences on hypoxic ventilatory control. In: Porter R, Knight J (eds) High altitude physiology: cardiac and respiratory aspects. Churchill, London, pp 89–97
Tetens V, Lykkeboe G (1981) Blood respiratory properties of rainbow trout, *Salmo gairdneri:* responses to hypoxia acclimation and anoxic incubation of blood in vitro. J Comp Physiol 145:117–125
Thyberg J, Sierakowska H, Edström JE, Burvall K, Pigon A (1982) Mitochondrial distribution and ATP levels in Chironomus salivary gland cells as related to growth, metabolic activity, and atmospheric oxygen tension. Dev Biol 90:31–42
Timiras PS (1977) Hypoxia and the CNS: maturation and adaptations at high altitude. Int J Biometeor 21:147–156
Timiras PS, Hill R, Krum AA, Liss AW (1958) Carbohydrate metabolism in fed and fasted rats exposed to an altitude of 12,470 feet. Am J Physiol 193:415–424
Tipton KF (1972) Some properties of monoamine oxidase. In: Costa E, Sandler M (eds) Adv Biochem Psychopharm, vol V. Monoamine oxidases. Raven Press, New York, pp 11–24
Tolbert NE, Essner E (1981) Microbodies: peroxisomes and glyoxysomes. J Cell Biol 91:271s–283s
Torrance JD, Lenfant C, Cruz J, Marticorena E (1970/71) Oxygen transport mechanisms in residents at high altitude. Respir Physiol 11:1–15
Torrance RW (ed) (1966) Arterial chemoreceptors. Blackwell, Oxford, p 402
Trojan S, Stipek S, Crkovska J, Martinek J (1978) Effect of repeated altitude hypoxia on the nucleic acid content of the rat prosencephalon. Physiol Bohemoslov 27:91–93

Tucker A, Horvath SM (1974) Regional blood flow responses to hypoxia and exercise in altitude-adapted rats. Eur J Appl Physiol 33:139–150

Tucker A, McMurtry IF, Reeves JT, Alexander AF, Will DH, Grover RF (1975) Lung vascular smooth muscle as determinant of pulmonary hypertension at altitude. Am J Physiol 228:762–767

Turek Z, Frans A, Kreuzer F (1972a) Hypoxic pulmonary steady-state diffusing capacity for CO and alveolar-arterial O_2 pressure differences in growing rats after adaptation to a simulated altitude of 3500 m. Pflügers Arch 335:1–9

Turek Z, Grandtner M, Kreuzer F (1972b) Cardiac hypertrophy, capillary and muscle fiber density, muscle fiber diameter, capillary radius and diffusion distance in the myocardium of growing rats adapted to a simulated altitude of 3500 m. Pflügers Arch 335:19–28

Turek Z, Kreuzer F, Hoofd LJC (1973a) Advantage or disadvantage of a decrease of blood oxygen affinity for tissue oxygen supply at hypoxia. A theoretical study comparing man and rat. Pflügers Arch 342:185–197

Turek Z, Ringnalda BEM, Grandtner M, Kreuzer F (1973b) Myoglobin distribution in the heart of growing rats exposed to a simulated altitude of 3500 m in their youth or born in the low pressure chamber. Pflügers Arch 340:1–40

Turek Z, Claessens R, Ringnalda BEM, Kreuzer F (1978a) Blood volume and body haematocrit of rats native to a simulated altitude of 3500 m. Pflügers Arch 374:285–288

Turek Z, Kreuzer F, Ringnalda BEM (1978b) Blood gases at several levels of oxygenation in rats with a leftshifted blood oxygen dissociation curve. Pflügers Arch 376:7–13

Turek Z, Kreuzer F, Turek-Maischeider M, Ringnalda BEM (1978c) Blood O_2 content, cariac output, and flow to organs at several levels of oxygenation in rats with a left-shifted blood oxygen dissociation curve. Pflügers Arch 376:201–207

Turek Z, Ringnalda BEM, Moran O, Kreuzer F (1980) Oxygen transport in guinea pigs native to high altitude (Junin, Peru, 4,105 m). Pflügers Arch 384:109–115

Ungar A, Bouverot P (1980) The ventilatory responses of conscious dogs to isocapnic oxygen tests. A method of exploring the central component of respiratory drive and its dependence on O_2 and CO_2. Respir Physiol 39:183–197

Valdivia E (1958) Total capillary bed in striated muscle of guinea pigs native to the Peruvian mountains. Am J Physiol 194:585–589

Vaughan BE, Pace N (1956) Changes in myoglobin content of the high altitude acclimatized rat. Am J Physiol 185:549–556

Velásquez T (1959) Tolerance to acute anoxia in high altitude natives. J Appl Physiol 14:357–362

Velásquez T, Martinez C, Pezzia W, Gallardo N (1968) Ventilatory effects of oxygen in high altitude natives. Respir Physiol 5:211–220

Verbrugghe C, Laurent P, Bouverot P (1982) Chemoreflex drive of ventilation in the awake miniature pig. Respir Physiol 47:379–391

Vergnes H (1973) Le métabolisme énergétique myocardique chez le rat vivant en haute altitude. Ph D Thes, Toulouse, p 162

Vergnes H, Moret P, Duchosal F (1976) Changes in biochemical properties of myocardial lactate dehydrogenase during exposure of rats to high altitude. Enzyme 21:66–75

Visschedijk AHJ, Rahn H (1981) Incubation of chicken eggs at altitude: theoretical consideration of optimal gas composition. Br Poultry Sci 22:451–460

Visschedijk AHJ, Ar A, Rahn H, Piiper J (1980) The independent effects of atmospheric pressure and oxygen partial pressure on gas exchange of the chicken embryo. Respir Physiol 39:33–44

Vogel JA, Hansen JE, Harris CW (1967) Cardiovascular responses in man during exhaustive work at sea level and high altitude. J Appl Physiol 23:531–539

Vogel JA, Pulver RI, Burton TM (1969) Regional blood flow distribution during simulated high-altitude exposure. Fed Proc 28:1155–1159

Vogel JA, Hartley LH, Cruz JC, Hogan RP (1974a) Cardiac output during exercise in sea level residents at sea level and high altitude. J Appl Physiol 36:169–172

Vogel JA, Hartley LH, Cruz JC (1974b) Cardiac output during exercise in altitude natives at sea level and high altitude. J Appl Physiol 36:173–176

Wade JG, Larson CP, Hickey RF, Ehrenfeld WK, Severinghaus JW (1970) Effect of carotid endarterectomy on carotid chemoreceptor and baroreceptor function in man. New Engl J Med 282:823–829

Wagner PD (1977) Diffusion and chemical reaction in pulmonary gas exchange. Physiol Rev 57:257–312

Walker E, Warnick F, Hamlet S, Lange K, Davis M, Uible H, Wright P, Paradiso J (1975) Mammals of the world, 3rd edn. Hopkins, Baltimore

Walker JEC, Wells RE, Merrill EW (1961) Heat and water exchange in the respiratory tract. Am J Med 30:259–267

Wangensteen OD (1972) Gas exchange by a bird's embryo. Respir Physiol 14:64–74

Wangensteen OD, Rahn H (1970/71) Respiratory gas exchange by the avian embryo. Respir Physiol 11:31–45

Wangensteen OD, Wilson D, Rahn H (1970/71) Diffusion of gases across the shell of the hen's egg. Respir Physiol 11:16–30

Wangensteen OD, Rahn H, Burton RR, Smith AH (1974) Respiratory gas exchange of high altitude adapted chick embryos. Respir Physiol 21:61–70

Watt JG, Dumke PR, Comroe JH (1943) Effects of inhalation of 100 percent and 14 percent oxygen upon respiration of unanesthetized dogs before and after chemoreceptor denervation. Am J Physiol 138:610–617

Weathers WW, Snyder GK (1974) Functional acclimation of Japanese quail to simulated high-altitude. J Comp Physiol 93:127–137

Webber PJ (ed) (1979) High altitude geoecology. Westview, Boulder, Col, p 188

Weber RE, Lykkeboe G (1978) Respiratory adaptations in carp blood. Influences of hypoxia, red cell organic phosphates, divalent cations and CO_2 on hemoglobin-oxygen affinity. J Comp Physiol 128:127–137

Wegener G (1981) Comparative aspects of energy metabolism in nonmammalian brains under normoxic and hypoxic conditions. In: Stefanovitch V (ed) Animal models and hypoxia. Pergamon Press, Oxford New York, pp 87–107

Weibel ER (1973) Morphological basis of alveolar capillary gas exchange. Physiol Rev 53:419–495

Weibel ER (1979) Oxygen demand and the size of respiratory structures in mammals. In: Wood SC, Lenfant C (eds) Evolution of respiratory processes. Dekker, New York, pp 289–346

Weihe WH (ed) (1964) The physiological effects of high altitude. Pergamon Press, Oxford New York, p 351

Weil JV, Zwillich CW (1976) Assessment of ventilatory response to hypoxia. Methods and interpretation. Chest 70 (1 Suppl):124–128

Weil JV, Battock DJ, Grover RF, Chidsey CA (1969) Venoconstriction in man upon ascent to high altitude: studies on potential mechanisms. Fed Proc 28:1160–1163

Weil JV, Kryger MH, Scoggin CH (1978) Sleep and breathing at high altitude. In: Guilleminault C, Dement WC (eds) Sleep apnea syndromes. Liss, New York, pp 119–135

Weiskopf RB, Severinghaus JW (1972) Lack of effect of high altitude on hemoglobin oxygen affinity. J Appl Physiol 33:276–277

Weiskopf RB, Gabel RA, Fencl V (1976) Alkaline shift in lumbar and intracranial CSF in man after 5 days at high altitude. J Appl Physiol 41:93–97

Weiss HS (1978) Role of shell diffusion area in incubating eggs at simulated high altitude. J Appl Physiol: Respir Environm Ex Physiol 45:551–556

West JB (1962) Diffusing capacity of the lung for carbon monoxide at high altitude. J Appl Physiol 17:421–426

West JB (1969) Ventilation-perfusion inequality and overall gas exchange in computer models of the lung. Respir Physiol 7:88–110

West JB (1977) Regional differences in the lung. Academic Press, London New York, p 488

West JB (ed) (1981) High altitude physiology. Hutchinson Ross Publ, Stroudsburg, Pa, p 461

West JB (1982) Respiratory and circulatory control at high altitudes. J Exp Biol 100:147–157

West JB (1983) Climbing Mt. Everest without oxygen: An analysis of maximal exercise during extreme hypoxia. Respir Physiol 52:265–279

West JB, Wagner PD (1980) Predicted gas exchange on the summit of Mt. Everest. Respir Physiol 42:1–16

West JB, Lahiri S, Gill MB, Milledge JS, Pugh LGCE, Ward MP (1962) Arterial oxygen saturation during exercise at high altitude. J Appl Physiol 17:617–621

Westerterp K (1977) How rats economize. Energy loss in starvation. Physiol Zool 50:331–362

Weyne J, Leuven F van, Leusen I (1977) Brain aminoacids in conscious rats in chronic normocapnic and hypocapnic hypoxemia. Respir Physiol 31:231–239

White FN (1978) Comparative aspects of vertebrate cardiorespiratory physiology. Annu Rev Physiol 40:471–499

White FN, Somero G (1982) Acid-base regulation and phospholipid adaptations to temperature: time courses and physiological significance of modifying the milieu for protein function. Physiol Rev 62:40–90

Widdicombe JG (1981) Nervous receptors in the respiratory tract and lungs. In: Hornbein TF (ed) Regulation of breathing, part 1. Dekker, New York, pp 429–472

Wikström M, Krab K, Saraste M (1981) Cytochrome oxidase. A synthesis. Academic Press, London New York, p 198

Will DH, Hicks JL, Card CS, Alexander AF (1975) Inherited susceptibility of cattle to high-altitude pulmonary hypertension. J Appl Physiol 38:491–494

Williamson DH, Lund P, Krebs HA (1967) The redox state of free nicotinamide-adenine dinucleotide in the cytoplasm and mitochondria of rat liver. Biochem J 103:514–527

Wilson DF, Erecinska M, Brown C, Silver IA (1977 a) Effect of oxygen tension on cellular energetics. Am J Physiol 233 (Cell Physiol 2):C135–C140

Wilson DF, Owen CS, Holian A (1977 b) Control of mitochondrial respiration: a quantitative evaluation of the roles of cytochrome c and oxygen. Arch Biochem Biophys 182:749–762

Wilson DF, Erecinska M, Drown C, Silver IA (1979 a) The oxygen dependence of cellular energy metabolism. Arch Biochem Biophys 195:485–493

Wilson DF, Owen CS, Erecinska M (1979 b) Quantitative dependence of mitochondrial oxidative phosphorylation on oxygen concentration: a mathematical model. Arch Biochem Biophys 195:494–504

Wilson DF, Erecinska M, Silver IA, Owen CS (1980) Metabolic sensing of cellular oxygen tension. Proc XXVIII IUPS Congr, Budapest, 14:280

Winn HR, Rubio R, Berne RM (1981) Brain adenosine concentration during hypoxia in rats. Am J Physiol 241 (Heart Circ Physiol 10): H235–H242

Winslow RM, Morrissey JM, Berger RL, Smith PD, Gibson CC (1978) Variability of oxygen affinity of normal blood: an automated method of measurement. J Appl Physiol: Respir Environ Ex Physiol 45:289–297

Winslow RM, Statham B, Gibson C, Dixon E, Monge C, Moran O, Debrot J (1979) Improved oxygen delivery after phlebotomy in polycythemic natives of high altitude. Blood 54 (Suppl 1):61A

Winslow RM, Monge CC, Statham NJ, Gibson CG, Charache S, Whittembury J, Moran O, Berger RL (1981) Variability of oxygen affinity of blood: human subjects native to high altitude. J Appl Physiol: Respir Environ Ex Physiol 51:1411–1416

Wittenberg BA, Wittenberg JB, Caldwell PRB (1975) Role of myoglobin in the oxygen supply to red skeletal muscle. J Biol Chem 250:9038–9043

Wittenberg JB (1970) Myoglobin-facilitated oxygen diffusion: role of myoglobin in oxygen entry into muscle. Physiol Rev 50:559–636

Wittenberg JB, Wittenberg BA (1981) Facilitated oxygen diffusion by oxygen carriers. In: Gilbert DL (ed) Oxygen and living processes. Springer, Berlin Heidelberg New York, pp 177–199

Wood CM, Shelton G (1980) The reflex control of heart rate and cardiac output in the rainbow trout: interactive influence of hypoxia, haemorrhage, and systemic vasomotor tone. J Exp Biol 87:271–284

Wood SC (1980) Adaptation of red blood cell function to hypoxia and temperature in ectothermic vertebrates. Am Zool 20:163–172

Wood SC, Johansen K (1972) Adaptation to hypoxia by increased HbO_2 affinity and decreased red cell ATP concentration. Nature (London) 237:278–279

Wood SC, Lenfant CJM (1976) Respiration: mechanics, control, and gas exchange. In: Gans C (ed) Biology of the reptilia, vol V. Academic Press, London New York, pp 225–274

Wood SC, Lenfant C (1979) Oxygen transport and oxygen delivery. In: Wood SC, Lenfant C (eds) Evolution of respiratory processes; a comparative approach. Dekker, New York, pp 193–223

Wulff LY, Braden IA, Shillito FH, Tomashefski JF (1968) Physiological factors relating to terrestrial altitudes: A bibliography. Ohio State Univ Press, RF Project 2360

Yousef MK, Horvath SM, Bullard RW (eds) (1972) Physiological adaptations: desert and mountains. Academic Press, London New York, p 258

Ziegler FD (1967) Respiratory and phosphorylative responses to hypoxia and food restriction. Am J Physiol 212:197–202

Zuntz N, Loewy A, Muller F, Caspari W (1906) Höhenklima und Bergwanderungen. Dtsch Verlagsh, Berlin, p 38

Zutshi DP, Kaul V, Vass KK (1972) Limnology of high altitude Kashmir lakes. Verh Int Verein Limnol 18:599–604

Subject Index

Volume 8: **W. Leuthold**

African Ungulates

A Comparative Review of Their Ethology and Behavioral Ecology

1977. 55 figures, 7 tables. XIII, 307 pages.
ISBN 3-540-07951-3

·...Dr. Leuthold displays a masterly command of his subject.... The work is basically a review of published knowledge with an original approach, enlivened by the author's interpretations and based on his intimate first-hand knowledge of the subject. The first chapter, on the application of ethological knowledge to wildlife management, covers an important area... a wealth of references is given so that the chapter provides useful guide to the literature. the illustrations are good and well chosen to demonstrate points made verbally and not first to embellish the text. The book will provide **excellent background reading for undergraduates and research students as well as for anyone seriously interested in African wildlife.** On the whole, **it can be thoroughly recommended."**

J. Applied Ecology

Volume 9: **E. B. Edney**

Water Balance in Land Arthropods

1977. 109 figures, 36 tables. XII, 282 pages.
ISBN 3-540-08084-8

... Dr. Erdney has provided a wealth of organized information on prior work and ideas for needed research, **all of which make the book a bargain.** The volume should prove useful, not only to those who work in arthopod water relations (it is a must for them), but also those of us interested in invertebrate and general ecology, entomology, comparative physiology, and biophysics." *AWRA Water Res. Bull.*

Volume 10: **H.-U. Thiele**

Carabid Beetles in Their Environments

A Study on Habitat Selection by Adaptations in Physiology and Behaviour

Translated from the German by J. Wieser
1977. 152 figures, 58 tables. XVII, 369 pages.
ISBN 3-540-08306-5

"...Because the book is comparative both in method and interpretation, it is a contribution to systematics as well as to ecology... **a fine synthesis of current knowledge** of homeostatic aspects of ecological relationships of carabids, and it is a fitting tribute to the man to whom it is dedicated: Carl H. Lindroth, who was instrumental in formulating the approaches and techniques that are commonly used in ecological research on these fine beetles. The materials is **well organized** and the text is **easily readable, thanks to the clarity of thought and expression** of the author and to the skill of an able translator." *Science*

Volume 11: **M. H. A. Keenleyside**

Diversity and Adaptation in Fish Behaviour

1979. 67 figures, 15 tables. XIII, 208 pages.
ISBN 3-540-09587-X

"... it is important as the first serious attempt by a senior researcher to produce an overview of the discipline. Previous works have all been symposium volumes or collections of papers haphazardly assembled, and Keenleyside has produced a volume that is of substantially greater value than these. In clearly perceiving that the unique and valuable features of fish behavior are its diversity of form and circumstance, he has charted a course that future authors would be wise to follow.
The book is well produced, well written, and easy to read. The illustrations are clear and straightforward." *Science*

Volume 12: **E. Skadhauge**

Osmoregulation in Birds

1981. 42 figures. X, 203 pages. ISBN 3-540-10546-8

Contents: Introduction. – Intake of Water and Sodium Chloride. – Uptake Through the Gut. – Evaporation. – Function of the Kidney. – Function of the Cloaca. – Function of the Salt Gland. – Interaction Among the Excretory Organs. – A Brief Survey of Hormones and Osmoregulation. – Problems of Life in the Desert, of Migration, and of Egg-Laying. – References. – Systematic and Species Index. – Subject Index.

Volume 13: **S. Nilsson**

Autonomic Nerve Function in the Vertebrates

1983. 83 figures. XIV, 253 pages. ISBN 3-540-12124-2

Contents: Introduction. – Anatomy of the Vertebrate Autonomic Nervous Systems. – Neurotransmission. – Receptors for Transmitter Substances. – Chemical Tools. – Chromaffin Tissue. – The Circulatory System. – Spleen. – The Alimentary Canal. – Swimbladder and Lung. – Urinary Bladder. – Iris. – Chromatophores. – Concluding Remarks. – References. – Subject Index.

Volume 14: **A.D. Hasler, A.T. Scholz**

Olfactory Imprinting and Homing in Salmon

Investigations into the Mechanism of the Imprinting Process

In collaboration with R. W. Goy
1983. 25 figures. XIX, 134 pages. ISBN 3-540-12519-1

Contents: Olfactory Imprinting and Homing in Salmon: Notes on the Life History of Coho Salmon. Imprinting to Olfactory Cues: The Basis for Home-Stream Selection by Salmon. – Hormonal Regulation of Smolt Transformation and Olfactory Imprinting in Salmon: Factors Influencing Smolt Transformation: Effects of Seasonal Fluctuations in Hormone Levels on Transitions in Morphology, Physiology, and Behavior. Fluctuations in Hormone Levels During the Spawning Migration: Effects on Olfactory Sensitivity to Imprinted Odors. Thyroid Activation of Olfactory Imprinting in Coho Salmon. Endogenous and Environmental Control of Smolt Transformation. – Postscript. – References. – Subject Index.

Springer-Verlag
Berlin
Heidelberg
New York
Tokyo